From Systems to Actor-Networks

A Paradigm Shift in the Social Sciences

By

Andréa Belliger and David J. Krieger

From Systems to Actor-Networks: A Paradigm Shift in the Social Sciences

By Andréa Belliger and David J. Krieger

This book first published 2024

Ethics International Press Ltd, UK

British Library Cataloguing in Publication Data

A catalogue record for this book is available from the British Library

Print Book ISBN: 978-1-80441-336-4

eBook ISBN: 978-1-80441-337-1

Contents

Introduction

This book attempts to document a tectonic shift in world views, or what can be called a "paradigm shift," not only in the sciences but also in society. There are many indications that in our time, a particular way of understanding the world, which in its depth and breadth could be compared to a tectonic plate, is by a process of historical and cultural subduction being pushed beneath another. A new way of seeing the world is consequently rising to the surface. These two tectonic plates are continents of knowledge, values, basic assumptions, and methods and, for this reason, can be called scientific paradigms. The two paradigms we will be concerned with in this book are the "systems paradigm" and the "network paradigm." Systems are everywhere, or at least they used to be. Now we are seeing that where systems once were, networks are appearing. Systems and networks are two different but related forms of order. Order in all its forms is what science investigates and attempts to understand. Striving for order is at the basis of societies, cultures, ideologies, and world views. Science is part of this program. We will argue that there are three levels upon which order appears: 1) the physical level of matter, energy, particles, and fields, which is the object domain of physics, chemistry, and related disciplines; 2) the order of life, which is studied by biology; and 3) the order of meaning which is described by philosophy and the human and social sciences. A scientific theory becomes a paradigm when it proposes to model order on all three levels. A paradigm is, therefore, a general or *universal* theory because it claims not only to explain one domain of reality but all of reality, including thereby – and this is very important – itself. Physics may claim to be a Theory of Everything, but it can neither account for life nor for the observer, that is, for itself, at least not yet. The same can be said of biology. Where in the brain, we may ask, is the observer to be found? We find neurons and patterns of their interaction, but we do not see the observer. We see, for example, correlations between the firing patterns of neurons and the vision of a rose. However, correlation is not causation and indeed not identity. Where is the similarity, sameness, identity of patterns of firing neurons, and the idea of a rose? This leaves the human and social sciences, which do explicitly study themselves. Their problem is to describe the kind of order they represent. Traditionally, philosophy and the human sciences have located their object domain on the level of meaning, albeit under many different names and mostly in opposition to nature. When these sciences include basic concepts and models applicable on the physical and biological levels, they become universal theories that claim to explain everything, including themselves.

In recent times, two general theories claim universality in that they bring together concepts and methods that are claimed to be applicable on all three levels of emergent order, the physical, biological, and the level of meaning. These are general systems theory and network theory. For general systems theory, all forms of order are systemic, from fundamental particles to living beings, human culture and society, and even the world (Earth, Gaia, Cosmos). Ecology, as we will see, has become a title for universal theories, and both systems theory and network theory offer their own "solutions" to the problem of a holistic understanding with which the Anthropocene confronts us. For systems theory, at least since the founding of cybernetics in the 1940s, it has become clear that both nature and human society are forms of systemic order. Network theory, at least actor-network theory, which we will examine in this book, explicitly rejects the typically modern distinctions between nature and society, subject and object, mind and reality. This implies that when speaking of meaning, one is speaking of being on all levels. According to these two theories, whenever we find order, whether in matter, life, or meaning, what we see are systems of one kind or another or networks in various forms. General systems theory, as opposed to the various special forms of systems analysis, is "general" because it describes systemic order as such, the general principles of any kind of systemic order, no matter whether physical, biological, or social. This is what allows it to claim the status of a paradigm. Actor-network theory also strives for a level of generality that encompasses everything. It does this by attempting to describe the mechanisms or processes that construct networks as such, no matter how small or large and no matter what actors, whether humans or nonhumans or both, are linked together. The principles of network order are claimed to apply on the physical, biological, and social levels, mainly because these distinctions are simply rejected. Whereas systems analysis emphasizes the reduction of complexity, holism, teleology, control, and stability, network thinking emphasizes connectivity, flow, participation, flexibility, and other values which we will discuss below that systems do not exhibit.

This book examines the claims of these theories critically and argues that we are witnessing a paradigm shift from systems to networks in our approach to science, society, and ourselves. This may seem a daunting task, and it indeed is, but the focus on generality unburdens us of the necessity of delving into details of systems or network analysis in the many different disciplines in which they are to be found. In a way, describing a paradigm is much easier than applying it. This means that the heroes of our story are theories, their basic concepts, their coherence, heuristic power, and their explanatory scope. The two thinkers that we will take as representatives of the theories discussed in this book are Niklas Luhmann for systems theory and Bruno Latour for actor-network theory. However, even if we let Luhmann and Latour do most of the talking, this is not a book about Luhmann or Latour or both together. It is a book about theory construction. The goal is to construct at least the outlines of a theory that answers to the needs of our time, and we are simply using what, in our opinion, are the best available resources. Although we talk about the social sciences, as do both the theories we will be examining; we are not doing social science in any traditional sense of the term.[1] We will accordingly not be primarily concerned with the many different forms of social systems or with the various networks making up the "collective," which is Latour's word for society. Instead, our focus is on the principles of systemic order as such and correspondingly on the general mechanisms of network construction. We will be concerned with the question of whether meaning can be successfully modeled as a system and, if not, whether the network model could be more successful. Posing this same question somewhat differently, we ask whether systems theory can explain itself and, if not, whether the network model offers a more adequate answer.

We begin in Part 1 by examining a theory that has several names, cybernetics, the theory of self-organization, complexity theory, constructivism,

---

1 Both Luhmann and Latour reject the traditional self-understanding of the social sciences.

or, as it is commonly known, systems theory.[2] Whatever name it goes by, we are dealing with an interdisciplinary and self-proclaimed universal theory that seeks to "explain" order not only on the physical and biological levels but also psychological and social phenomena. Just as a spider's web is fixed on all sides at different and often widely separated points, so today what could be called "general systems theory" is based on different and quite distant disciplines: cybernetics and computer science, cognitive science, physics, engineering, biology, logic and mathematics, psychology, neurophysiology, anthropology, sociology, semiotics, organization theory, and philosophy are among the sources of systems models and concepts. When referring to a general systems theory, we must keep in mind that because of these many sources and applications, we are dealing much more with a "paradigm" (László, 1974) than with a unified theory with established methods and particular domains of application. The systems paradigm consists of a more or less closely associated set of basic concepts, terms, approaches to phenomena, methods, and beyond this, a worldview that understands all forms of order as in some way "systemic." Whenever one speaks of systems, whether in physics, chemistry, biology, psychology, sociology, economics, computer science, engineering, neuroscience, management studies, etc., one usually finds the associated concepts of self-organization, complexity, adaptation, self-regulation, learning, information, feedback, holism, teleonomy, control, homeostasis, etc.

Despite the tendency to use similar concepts in many different disciplines, which is typical of a scientific paradigm, general systems theory "...cannot currently be presented as a consolidated set of basic concepts, axioms,

---

[2] The idea of an interdisciplinary, universal research program under the name of systems theory originated with Ludwig von Bertalanffy. Bertalanffy's systems theory, together with Shannon and Weaver's (1949) information theory and Wiener's (1948) cybernetics, Maturana and Varela's theory of autopoiesis, as well as Bateson's work in psychology and anthropology, and Parsons's sociology, Beer in organization theory, and more recently theories in biophysics, cognitive science, artificial intelligence and associated disciplines form a model, basic concepts, and methods that can be found in a wide variety of scientific disciplines. Even though cybernetics and systems theory are not current buzzwords, their influence is everywhere, so it is justified to speak of a systems paradigm in the sciences and beyond science in our current cultural vernacular. Since there are many kinds of systems on the physical, biological, and human levels, we will speak of systems (plural) instead of system (singular).

and derived propositions" (Luhmann, 1984, p. 34). This is because the systems theory research program is not a specific discipline but rather a discourse, i.e., a particular way of talking, wherein the most diverse topics are approached using similar terms and similar assumptions. Therefore, the question arises: "...on which common or at least related conceptual foundations this transdisciplinary cooperation is based" (Krohn/Küppers/Paslack 1987:459).[3] For example, what do we mean when we talk about feedback loops, reduction of complexity, autopoiesis, information construction, codes, self-organization, homeostasis, adaptation, evolution, sustainability, etc.? To what objects and processes can we legitimately apply such terms? Do these terms have a common meaning when applied to very different phenomena, or are they mere equivocations? To what extent is it useful or even possible to transfer physical and biological models to the humanities and talk about cognitive, psychological, and social systems, or the other way around, to speak of physical systems as information processing and life as cognition? And if this is common practice, on what explicit and implicit presuppositions is this based? The self-evidence with which systems theoretical concepts are used in psychology, sociology, anthropology, philosophy, business, politics, education, and indeed everywhere challenges us to ask these and similar questions.[4] The attempt to present an overview of the models and basic concepts of systems theory thus seems to be the order of the day, and this is especially so when systems theory actually functions like a "paradigm," informing, implicitly if not explicitly, all aspects of life and society.

Our discussion of the systems paradigm does not intend to trace the historical lines of development of the various systems theoretical approaches in the sciences, nor to describe the various versions of systems

---

3 All translations from German language publications listed as such in the Literature are by the authors.

4 On the ubiquity of systems thinking and assumptions, see, for example, the self-evident value of "sustainability" as exemplified by the WHO Sustainability Development Goals, which claim to "transform the world" (https://www.who.int/europe/about-us/our-work/sustainable-development-goals). Also noteworthy in this regard is the new critique of "environmentalism" and the idea of a "General Ecology" (see Hörl/Burton 2017), and finally, there looms the prospect of a data-driven society governed by algorithms (see Kalpokas 2019; Schuilenberg/Peeters 2020 for an overview).

theory currently in use,[5] but rather to attempt to set forth the conceptual framework or the conceptual "common property" of systems theoretical discourse. It should be emphasized that this is done from a particular perspective. We are guided by the question of the extent to which systems theory can successfully be applied on the specifically human level of order, that is the level of meaning. We will not be primarily concerned with physical or mechanical or biological systems but with meaning systems. Our focus will be on the level of emergent order that can be called meaning, that is, the form of order that philosophy and the social sciences study and to which they also belong. This is, therefore, not a book on physics, biology, cognitive science, or artificial intelligence, even though we will have occasion to refer to these disciplines. Our concern is with the extent to which order on the level of meaning can be understood as systemic. And if this turns out not to be the case, what theoretical model could do a better job? This is where the comparison of systems to networks and the proposals of the new network paradigm become relevant.

The justification of prioritizing meaning over matter and life lies in the fact that whatever we know about matter and whatever we know about self-organizing, adaptive systems, otherwise known as life, we know this on the level of meaning and as forms of meaning. As Heidegger never tired of pointing out, what is closest to us and what comes first is not knowledge of objective facts that science presents but the world of meaning that includes all aspects of our existence. Science, after all, is a social phenomenon, part of our social practices, and not the other way around, and this remains so even if society becomes the object of scientific investigation. If everything we can know and be, including all the different sciences, is primarily ordered on the level of meaning, then

---

[5] This has been done by some of the principle actors themselves, various organizations, as well as later commentators. For an overview, see the Wikipedia article on System Theory https://en.wikipedia.org/wiki/Systems_theory#cite_note-:1-5 . For historical descriptions, see the website of the American Association for Cybernetics https://asc-cybernetics.org/foundations/history.htm; also, the website of Principia Cybernetica Web http://pespmc1.vub.ac.be/CYBSHIST.html. And the website of the International Society for Systems Science https://www.isss.org/home/. Recent monographs include Malapi-Nelson (2017) and the work of Bruce Clark (2019). Short histories have been offered by Umpleby (see papers at: https://blogs.gwu.edu/umpleby/papers-on-cybernetics-and-systems-science/; 2008) and Adams et al. (2013).

it is on this level that any claims for the universality of a theory must be justified. It must be shown that a theory that claims to be universal can explain meaning on its own terms and not reduce it to matter or life. A theory that cannot adequately explain itself cannot claim to be a universal theory. No matter how much of reality it covers, it leaves something out, namely, itself. Taking this into account requires that we pay more attention in this book to philosophical and sociological systems theory than physical or biological versions of the same. Since the most prolific and influential sociological systems theorist is Niklas Luhmann, we will be following the exposition of systems theory in his work as our primary source.[6] Following Luhmann, we will focus on the idea of *communication* as a system. In emphasizing communication, Luhmann is not alone. For many authors and in many respects – see the linguistic turn in philosophy, semiotics, structuralism, and poststructuralism, and Habermas' *Theory of Communicative Action* – the constitution of meaning in linguistic (also non-verbal!) communication has acted as an "attractor" for the study of human existence in society.

The fruitfulness, heuristic power, and above all, the universality of systems theoretical models and concepts have been demonstrated on the physical and biological levels, but can systems theory also successfully model meaning? How does meaning arise? What forms of order does meaning take on? Is meaning a systemic phenomenon or something else? As mentioned above, from today's scientific point of view, a theory can only fulfill its claim to universality if it can "explain" not only everything in the world but also itself. Physical, mechanistic, and biological theories notoriously omit the observer even while paying lip service to the need to explain who it is that is doing the science and building the machines in the first place. Where in the standard model is the observer? Which

---

6 None of the theories of 2nd order cybernetics that have arisen in the wake of Bateson and von Foerster in the English language come near to the theoretical sophistication and "systematicity" of Luhmann's work which incorporates their insights and goes beyond them. Furthermore, another reason for focusing on Luhmann's extensive work is that this work is not well-known in the English-speaking world. Very little of Luhmann's many writings have been translated, and even less discussion has been generated by his work. We hope with this book to contribute to making Luhmann better known outside the German-speaking world.

particle or field is it? Embarrassingly, the theory calls for the observer and even depends upon it but cannot explain it.[7] Perhaps the observer is a fundamental force of nature beyond the four known forces. If so, it has not been found, measured, or seen in the decay patterns of the LHC. Could it be that physics would need to look beyond matter to find itself? The same can be said of biology. Where is the observer in an organism? Is it the central nervous system? Can it be found in the brain? The fact that this has not been done and does not seem anywhere near being done, not even in the artificial brains devised by computer scientists, is called the "hard problem" of consciousness.[8] The problem is "hard" because it seems that the observer is not anything physical or biological at all. If one attempts to make the subject into an object, consciousness disappears, and if you try to know consciousness, it appears to be something that we do and not something we can know as an object. Neither physical nor biological science seems able to find anything like an observer beyond merely adaptive behavior in the objects they study. This means that they cannot explain their own activities, how they themselves are possible, or even what they are as scientists in society.[9]

Science itself is not a physical or biological phenomenon even if it admittedly takes place in some kind of still, despite claims to the contrary, inexplicable relation to physical and biological substrates. A substrate, whether it be matter with regard to life, or life with regard to meaning, is not a cause; it does not determine anything, and at the most, it constrains. What emerges out of it, on top of it, in relation to it, cannot be derived from it. This is why

---

[7] For a detailed discussion of the "nonseparation" of observer and observed in quantum physics and the ongoing unwillingness of scientists to correspondingly revise their basic notions of reality, see d'Espagnat (2006).

[8] See Wikipedia: https://en.wikipedia.org/wiki/Hard_problem_of_consciousness. Generally, the term "system" is used in physics to describe any object of study that consists of interrelated parts. The delimitation or designation of the object is done by the observer.

[9] Bateson (1979) notoriously proposed that mind and nature are one and emphasized the continuities between life and meaning as many after him have also done, for example, Michael Levin and Karl Friston who, following Maturana and Varela, equate life, that is, adaptive behavior, with cognition. Whether "cognition," as the term is used in cognitive science, can be equated with meaning is a question we will discuss below. As we shall see, it was left for Latour and actor-network theory to examine how science actually operates.

one speaks of *emergence*, that is, the appearance of something that cannot be deduced or reduced to what precedes it. On the one hand, the question of what science is must itself become a scientific question, while on the other hand, the science of science has yet to find a theoretical foundation since it can neither be derived from matter nor from life and can only be explained by an adequate theory of meaning. Does systems theory give us an adequate theory of meaning? Could it be that with general systems theory meaning finally succeeds in understanding itself, or is it not that with systems theory meaning misunderstands itself as a system? Considering that systems theory claims to be a universal theory, that is, a theory of everything, we must ask if systems theory can do what physics and biology cannot. We must ask: Can systems theory itself be understood and explained as a system? If systems theory cannot adequately explain itself, does this mean that systems theoretical concepts require further justification on an even more general level? If the discourse of systems theory wants to demonstrate its universality, then the epistemological and metaphysical presuppositions, that is, the implicit or explicit theory of the meaning of the systems theoretical paradigm, must be clarified. The observer must find itself in the theory and appear as a result of the theory. This was precisely the program of what has been called 2nd order cybernetics or constructivist epistemology.[10] Has this program been successful?

If it should turn out that systems theory cannot adequately model itself, and this will be the upshot of our argument in this book, then we must ask what can. This is where we introduce in Part 2 the network paradigm.[11] After examining Luhmann's theory of social systems in detail

[10] If one equates meaning with logic or conceptual coherence, as Luhmann in the end does, but Latour clearly does not, then the legacy of Hegel in the systems paradigm becomes apparent. For an overview of 2nd order cybernetics see Wikipedia https://en.wikipedia.org/wiki/Second-order_cybernetics#:~:text=More%20generally%2C%20second%2Dorder%20cybernetics,it%20is%20termed%20as%20such, and for constructivism: https://en.wikipedia.org/wiki/Radical_constructivism#:~:text=Radical%20constructivism%20is%20an%20approach,the%20world%20beyond%20that%20experience. Current influential proponents of this program are Levin (2022) and Friston (2010).

[11] For network science in physics and biology, see the work of Albert-Laszlo Barabasi (https://barabasi.com/), and for the social sciences, see the work of Bruno Latour (http://www.bruno-latour.fr/).

and attempting to show that there are many problems with the attempt to model meaning as a system, we turn to a competing scientific paradigm that claims to be able to do this. The network paradigm, just like the systems paradigm, is not a unified theory but a set of related concepts and approaches to reality which share a common worldview. This is a view of the world as being *networked*, that is, constructed by means of associations, links, relations, and, as we shall argue, information. Many promising developments use network models in physics, computer science, and artificial intelligence. Wolfram's multigraph[12] model for a physical theory of everything and Goetzel's OpenCog[13] model for AGI (artificial general intelligence) could well be cited as examples of the network paradigm. As interesting as these examples are, we will take our sociological focal point as a guide and rely on the work of Bruno Latour and what is known as actor-network theory to explicate the network paradigm. The focus will be on the social sciences and the theory of meaning.

Although Latour does not explicitly develop a theory of meaning, he clearly assumes that actor-networks are at once that of which the world consists and that this world is not to be understood according to the traditional, modern dichotomies of nature vs. society, subject vs. object, matter vs. mind, language vs. reality. Indeed, all the pillars of modern epistemology and ontology are pushed aside to make room for a close description of how humans and nonhumans enter into associations in order to construct the world, which Latour calls the "collective."[14] Since it consists of associations, links, or relations, the collective can be said to be "networked." Here is where actor-network theory is similar to what has come to be known as "network science." For the newly emerging science of networks, all forms of physical, biological, or social order exhibit network characteristics. They can be understood as relations or links between nodes. Where systems theory finds order in functional relations between the elements of a closed system, network theory finds order in the relations between

12 See https://writings.stephenwolfram.com/2020/04/finally-we-may-have-a-path-to-the-fundamental-theory-of-physics-and-its-beautiful/.

13 See https://en.wikipedia.org/wiki/OpenCog; https://opencog.org/.

14 The collective is everything that associations and relations gather together, and, as we shall see, this includes everything that exists.

actors in an open and flexible network. Where systems theory proclaims there are systems, network theory finds everywhere not systems but networks. We interpret Latour's "principle of irreduction"[15] namely, that nothing can be either reduced to another, or reduce another to itself, by which he describes the relational nature of all beings as a definition of *information*. We argue that actor networks are the results of *information construction*. This argument is based on a close analysis of the mechanisms of networking. Latour speaks of these mechanisms or processes in terms of "translating" and "enrolling" actors that are both human and nonhuman into networks. He also calls this "technical mediation" since meaning is always embodied in associations between humans and nonhumans, that is, things and artifacts. For actor-network theory, there is no disembodied, purely virtual, or ideal meaning. Meaning is Being. In short: If there is nothing in the hand, there is nothing in the head either. After describing how actor networks are constructed, we review Latour's description of the various forms of networks to be found in modern society. Both systems theory and actor-network theory offer empirical descriptions of society. For Luhmann, society consists of functional subsystems, whereas for Latour, society is a "collective" of "modes of existence."

In Part 3, we move beyond a purely theoretical comparison of systems theory and actor-network theory and attempt to empirically describe the shift from systems to networks in various areas of society. What does the advent of the network paradigm mean for society? We attempt to answer this question with reference to networking in business, education, and healthcare, which are representative for other areas of society as well. We describe how the guiding values arising from the affordances of digital technologies which dominate our historical moment lead to the transformation of systems into networks with regard to three characteristics: 1) As opposed to systems, networks do not define fixed roles and functions, 2) as opposed to systems, networks do not have constitutive boundaries, and 3) as opposed to systems, networks cannot be controlled top-down. A description of how these basic differences of systems to networks in the areas of business, education, and healthcare illustrate the emergence of a global network society.

---

[15] See Latour (1993).

Part I

General Systems Theory

Chapter 1

Basic Concepts of Systems Theory

1.1 In the Beginning, There Is the Difference

Let us start at the beginning, i.e., with nothingness. Let us imagine that there was only some kind of primal substrate at the beginning of the world. The original substrate is not differentiated and not formed. We can think of it according to the atomistic model as consisting of infinitely many uniform elements, which have complete freedom to form connections. Or we can use a distinction that Luhmann preferred, the distinction between medium and form, wherein the medium consists of "loosely coupled" (Luhmann, 2001) elements, and form arises when these elements become "closely coupled" *(verdichtet)* or in some way bound to each other such that order or a pattern arises and no longer everything is possible. In the primordial substrate, or medium, everything is the same. There are no differences. One cannot even distinguish among the elements since they are all the same, and there is no pattern or order, no relations binding them together in any particular way.[16]

If we now want to put ourselves in the role of God and create a world, then we must begin with a distinction. In the beginning, was the distinction, the difference. The idea of difference is a basic principle in systems theory. We can even speak of "difference theory" (Luhmann, 2006) since everything is based on a difference or a distinction. If there were no distinctions, there would be nothing. Identity, substance, and bonded unity is consequently not the main concept of systems theory.

---

[16] The fact that the primal substrate or medium consists of elements explains why differentiation is possible. If the original substrate is a single, massive "thing," then nothing could develop from it; internal differentiation and, thus, negentropy would be unthinkable. The paradox of undifferentiated differentiation could be overcome by asserting that the differentiation of the primal substrate contains no information. These are differences that make no difference. Order, on the contrary, is information. Other words for this substrate in systems theory are "complexity" and "entropy."

It is the boundary and not the thing that is more important. This implies that the primal elements, whatever they might be, cannot be thought of as some kind of things whose essential properties determine what the world can become. Order is relational and not substantial. In this respect, systems theory fundamentally differs from the Western philosophical tradition with its emphasis on substance or things. Identity is not what makes something to be what it is. Identity is not why something exists, its ground of being. In systems theory, something "is" only by being different from something else, i.e., by being a relation and not a thing. Systems ontology is not an ontology of things; it is a relational ontology.

Where differences in the form of relations and not mere multiplicity come from is a question we can leave open for now. Whether differences are arbitrarily introduced by a mythical creator god or a perhaps equally mythical "observer," whether differences arise in the evolutionary process of the self-organizing universe (Bateson, 1979; Maturana/Varela, 1987; Jantsch, 1980), whether they are historically culturally predetermined (Schmidt, 1994) or, finally, whether they are elements of a communication system specialized in the production of differences (Luhmann,1990b), these are questions that will be discussed below. For the moment, it is only important to note that systems are based on a distinction.

1.2 What is a System?

We now know what needs to be done if something is to be created from the primordial substrate: A distinction must be made. But how is a distinction to be made? A distinction is a delimitation. Something must be delimited. However, this is only possible if, to remain with the idea of an undifferentiated original substrate/medium in which all elements are equal, when some elements are *selected* and put together, that is, *related* in a certain way. Elements must be selected and related to each other or composed in a particular way. A composition of this kind is called in Greek *to systeme*.

A "system" is an ordered whole, something that has been put together in a particular way. By putting things together in a certain way, a distinction arises. It is not enough to just put some elements aside – where? – and it cannot be enough to build a wall around some elements – how do we imagine this if nothing is yet present in the world, not even walls? The parts of the original substrate, which are to be selected and related and thus distinguished from all other parts, must, to be distinguished, be ordered in a certain way. The parts must enter into certain relatively stable relations with each other. Otherwise, they are the same as all others, and nothing distinguishes them. Differentiation, therefore, takes place via *relationing*. Selection without relationing makes no sense and achieves nothing. However, selection and relationing, necessary as they are, are also not enough. The elements selected are related in a certain way such that something can be done, and some operation becomes possible. The parts are *steered* or *controlled* in order to operate to achieve some goal or purpose. A system, therefore, consists of processes of selecting, relationing, and steering/controlling. These are the fundamental principles of systemic order. Selection without relationing is no selection because the elements are not in any way distinguished from all others in the substrate. Relationing without controlling or steering operations has no purpose, and there would be no reason for anything to be related in any particular way.

To make a distinction that builds a system, we must 1) select some elements from the totality of everything and 2) relate these elements among themselves in a certain way, that is, 3) in a way that allows certain controllable operations. If one satisfies these three conditions, one has a system. A system, then, consists of elements that stand in certain relations to each other, which relations then enable certain operations or processes that pursue a goal due to control or steering of the elements and their relations. This *steering* function accounts for the close relationship between systems theory and cybernetics. The *Kubernetes* (Greek) was the helmsman of a ship, the one who steers the ship or also society. Therefore, the *Kubernetes* was also the governor. It is from this Greek word that the title cybernetics comes. In a cybernetic system, one speaks of a governor or controller that steers the system toward specific goals or setpoints. This third principle of steering explains why systems theory and cybernetics are the same. Cybernetics emphasizes the steering function, which, as we shall see, is associated with feedback loops, circular causality, information, and stability. Systems theory treats all three functions as equally important. One could therefore argue that systems theory is a broader and more encompassing term, whereas cybernetics emphasizes one albeit central aspect of systemic order. These three basic principles of *selection, relationing,* and *steering* will accompany us all the way and in every domain of systems theoretical discourse. They build the foundation of the epistemological and ontological presuppositions of systems theory. They are the basic principles of systemic order. However, as important as they are, there is more to general systems theory than these principles.

1.3 System and Environment

If one has created a system by a distinction that arises based on selection, relationing, and steering, the system must be different from something else. That which is different from the system, that which is "outside" the system, that which contains everything that does not belong to the system, and that with which the system in some way interacts is the *environment*. The system is fundamentally constituted by a *difference from the environment*. This is because the system necessarily includes its own elements and excludes everything else: the environment. The *system/environment difference* is at once inclusive and exclusive. The environment is everything that is not an element in the system and is under the control of the system and thus excluded from the system. The idea of selection implies that not everything can be included in the system. This has important consequences. It means there can be no environment in itself, and as a corollary, there can be no system of everything, no world system.[17] The system/environment difference is also a relation. The environment is always also the environment for or in relation to a system. Systems and environment necessarily imply

[17] We will return to this later when discussing the extent to which meaning and the idea of ecology, both of which include everything, can be modeled as systems.

each other, condition each other, and belong to each other. There is no system without an environment, and no environment without a system. In our systems theoretical world creation myth, the original substrate/medium cannot represent an environment before the world's creation, i.e., as long as there is no system that, by distinguishing itself, "makes" what is excluded into an environment. In this sense, one can probably speak of creation "out of nothing." Luhmann (1995:16) states:

> There is agreement within the discipline today that the point of departure for all systems-theoretical analysis must be the *difference between system and environment*. Systems are oriented by their environment, not just occasionally and adaptively, but structurally, and they cannot exist without an environment. They constitute and maintain themselves by generating and maintaining a difference from their environment, using their boundaries to regulate that difference. ... In this sense, *boundary* maintenance is system maintenance.

Moreover, concerning the environment, Luhmann (17) writes:

> The environment receives its unity through the system and only in relation to the system. It is delimited by open horizons, not by boundaries that can be crossed. Thus, it is not itself a system. It is different for every system because every system excludes only itself from its environment. Accordingly, the environment has no self-reflection or capacity to act.

1.4 Why Are There Systems? Reduction of Complexity (Negentropy)

Let us return to the primordial substrate or medium. The original substrate excludes nothing, so it cannot be a system. It is the "open horizon" containing "no boundaries that can be crossed." It is not a product of processes of selection, relationing, and steering. Accordingly, the number or quantity of possible elements in the original substrate is infinitely large, and also infinite is the number of possible connections among the elements. In the primeval substrate, everything is possible because every element can be connected with every other in all possible ways. This follows from the fact that in the primordial state, there are no given selections, relations, determinations, limits, or regulations that could influence whether this or that event would happen rather than any other event. It is impossible to say whether the elements would enter into this or that connection rather than any other. Such a primordial state, in which everything is possible, and everything is therefore equally probable, can be called chaos or, to take up a term from thermodynamics and information theory, *entropy*.

The original substrate is chaos or entropy, the equal probability of all possible relations. However, the primordial substrate can also be called *complex* because everything is possible. We can even say that it contains *absolute complexity*. If a distinction creates a system, then this system will necessarily be less complex than the initial state. This is because the system necessarily contains fewer elements than the environment. In addition, the system's elements are ordered in a certain way. Because of this, the system contains fewer possible relations among the elements than in the original substrate. In short, when a system comes into being, entropy is negated. Therefore, we can speak of *negentropy*, meaning complexity has been reduced. Systems always and necessarily reduce complexity.

> No system can realize the logical possibility of connecting every element to every other one. Systems necessarily reduce complexity. (Luhmann 1995:44).

In general, a system emerges as a reduction of complexity. Accordingly, complexity is the reason for the emergence of systems. Complexity presents itself as a *problem to be solved*. Because there is complexity, therefore there are systems. Suppose one wants to understand a system, i.e., explain the system by tracing it back to its ground. In that case, whether it is a machine, a living being, or a society, one finds at the beginning not a *causa sui* or a prime mover or a God, or even some primordial substrate, as we have been calling it, but rather a primordial problem, namely the problem of complexity. If systems emerge to reduce complexity, complexity is the problem that all systems, in one way or another, solve. Therefore, systems are dynamic problem solvers, and the problem they solve by means of selection, relationing, and steering is the problem of complexity.[18] Theoretically, it would be helpful to speak not of some primal substrate as we have done above. Instead, one could say that the environment represents a *problem space* in which systemic order arises as various proposed solutions. Solving a problem means that certain constraints have successfully been met. If the environment were chaotic, there would be no way of solving any problem because there would be no constraints in a purely random and chaotic space. For this reason, as we will see, systems always arise within an environment that is in some way already ordered. Therefore, the environment can be conceptualized as a set of constraints, however unknown and large this set may be. As problem solvers, systems can be more or less successful. They can solve the problem of complex environmental constraints well or less well or not at all and disappear. These considerations lead to ideas of adaptation, viability, and evolution and connect systems theory intimately to biology. As we shall see, most models of systemic order are taken from biology, and systems theoretical discourse is replete with biological metaphors.[19]

Admittedly, the notion of complexity is itself complex, and before we go further requires some explanation and clarification. First, it is important

---

[18] Systems are always functional entities, i.e., they "operate", they "do" something, and thus exist as "agents" that solve problems. The problem they solve on the most general level is always: How can complexity be reduced? As Luhmann (1995:25) puts it: "Establishing and maintaining the difference between a system and the environment then becomes the problem because, for each system, the environment is more complex than the system itself. Systems lack the 'requisite variety' (Ashby's term) that would enable them to react to every state of the environment, that is to say, to establish an environment exactly suited to the system. There is, in other words, no point-by-point correspondence between system and environment (such a condition would abolish the difference between system and environment). For this reason, establishing and maintaining this difference despite a difference in the degree of their relative complexities becomes the problem."

[19] One should recall that cybernetics and systems theory arose from the engineering problems of self-regulating machines (anti-aircraft guns steered by radar) and the purposeful behavior of organisms.

to note that complexity can only be reduced by complexity.[20] Selection, relationing, and steering not only exclude complexity but also create a complexity that is intrinsic or internal to the system. When one looks at the organization of elements in a system, the system appears to have its own complexity:

> We will call an interconnected collection of elements "complex" when, because of immanent constraints in the elements' connective capacity, it is no longer possible at any moment to connect every element with every other element. (Luhmann 1995:24)

What Luhmann here refers to as the "immanent constraints" of the connective capacity of the elements arises from the fact that a system always *constructs* its own elements to be elements of the system. Selection and relationing transform the original condition of the elements into system-specific elements. If we think of a table as a system, the elements of the table are the top and the legs. However, tops and legs do not lie around in the environment. The system constructs them, and once something, whether wood, metal, glass, etc., has become a tabletop or legs of a table, it is transformed, and its connective capacity has been constrained. We will return to this below.[21] What is important to note here is that selection and relationing at once limit and create complexity. They allow the system to have only certain states or operations. The more states a system can have, the more complex it can be said to be. And the more complex a system is, this was Ashby's law, the more it can reduce environmental complexity. Complexity, therefore, can paradoxically only be reduced by complexity. Because of "immanent constraints" on the connective capacity of the elements, absolute entropy, as the primal

---

[20] See Ashby's (1974:299) law of required diversity, "...only diversity can destroy diversity," which Luhmann cites in the previous note. For a recent theoretical formulation of this principle, see Assembly Theory proposed by Lee Cronin and Sarah Walker (2023).

[21] As the reader has probably noticed, the term "construct" and "construction" do heavy labor in this book. A precise definition is offered by Latour (2013:151ff.), where he argues that "construction" should not be understood as "artificial" in the sense that opposed to that which is "real." We will return to this later when discussing actor-network theory.

substrate characterizes it, is negated.[22] Not all connections between elements are possible. As already noted, one cannot conceive of the emergence of systems as a reduction of complexity without presupposing an environmental complexity, that is, as a set of constraints. Only under this condition can complexity be understood as a kind of motivation, the "cause," or the ground of system formation. Luhmann asserts that the environment is always more complex than the system and that, consequently, there is such a thing as "a totality of possible events," a "unit of reference that no longer has boundaries" (1970:115). Luhmann calls this all-encompassing, boundless field of complexity "world."[23] If we now remember that the environment itself is always system-relative, i.e., that there is no environment per se, but only the system's environment, then we have a third kind of complexity, namely the complexity of the system-relative environment. The environment is always the environment for a system relative to a system. In the end, we are talking about three kinds of complexity: 1) absolute complexity (the primal substrate, chaos/entropy), 2) the complexity of the environment relative to a particular system, the environment of the system, and 3) the internal complexity of the system itself. This complex situation can be represented as follows:

22 This is the point of departure for Karl Friston's Free Energy Principle and the application of Bayesian mechanics to systems theory.

23 On the concept of the world, see Luhmann (1970:115): "The world cannot be conceived as a system, because it has no 'outside' against which it demarcates itself. If one wanted to think of the world as a system, one would immediately have to think of an environment of the world, and the concept of the world guiding thinking would be displaced to this environment." The concept of "world" implies that complexity, although boundless, is nonetheless not chaos. See for Luhmann's concept of the world Günter Thomas (1992). We will return to this when discussing meaning as world. If meaning is the world, how can meaning be modeled as a system?

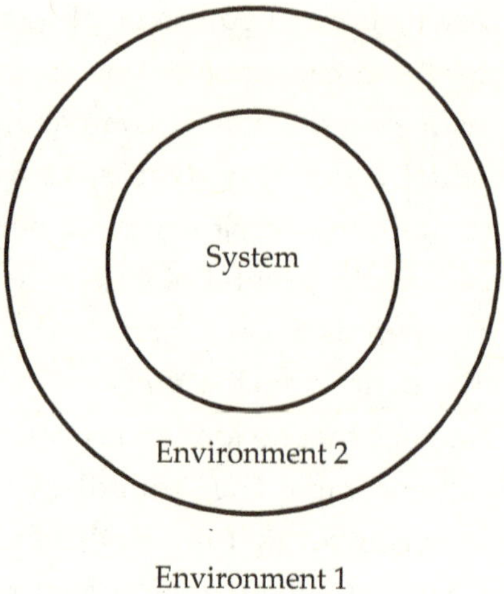

Figure 1.1

With respect to the first two forms of complexity, one could distinguish between "Environment-1" and "Environment-2." Environment-1 is the absolute complexity, which denotes the environment of all possible systems. As mentioned above, one could also consider it the problem of why anything should exist rather than nothing. Environment-2, on the other hand, would be the respective system-relative environment. This is what van Jakob von Uexküll referred to as *Umwelt* and which later became important in ecology and cognitive science. It is the world whose constraints a system must adapt to in order to exist. It is a problem that the system must solve. In biology, it can be called a "niche." Regarding meaning systems, as we shall see, the environment could be called a "cultural niche," which differs from biological or ecological environments not only in that it is constructed (see *Niche Construction Theory*[24]) but above all because it is included *within* the meaning system and not outside. We will return to this unique notion of environment that applies only to systems of meaning later. If one wants to use the idea of the "world" in this context, it is recommended that "world" refer to the system/environment difference specific to meaning systems. Metaphorically, of course, one could say that every organism has its own "world." The pond is the world of the frog that lives in it. The frog

---

[24] For an overview of the literature on NCT, see https://www.oxfordbibliographies.com/display/document/obo-9780199941728/obo-9780199941728-0089.xml.

has its own world, just as does any system capable of cognition. We will return to the distinction between environment-1 and environment-2 below in the discussion of the concept of meaning. We will see that the idea of two forms of the environment plays a significant role not only in living systems but also in human-level systems theory.

For the moment, it is important to note that the environment for the system cannot be absolutely entropic but must be thought of as always already ordered or reduced in some way.[25] The absolute complexity of a chaotic primordial substrate, i.e., a boundless entropy containing all possibilities, is not found in the real world. Moreover, it is also something in which no system could survive. It is a limit concept like the *materia prima* of ancient philosophy. It is only conceivable as the other of order, form, and structure, but in itself, it has no determinations or constraints. The necessity to talk about the indeterminate and to think of it as the difference between the determinate and determinable creates a certain fuzziness and ambiguity in the use of the concept of the environment. Environment, then, in the proper sense, is always already co-ordered by its own nature as well as systemic processes of selection, relationing, and steering:

> Every self-referential system has only the environmental contact it makes possible and no environment 'in itself.' But this 'itself makes possible' is not possible in a structureless, arbitrary, and chaotic environment because within such an environment, it is impossible to carry out "internally" satisfactory proof of worth and, from the perspective of evolution, to acquire permanence. (Luhmann 1995:101)

Systems, then, emerge within an environment that is always already ordered in a certain way and, in this sense, "complex." The (pre-) ordered environment sets constraints or conditions of possibility for a system insofar as the elements available to the system are already

[25] Mathematically, the relations between the system and environment can be modeled as "Markov blankets," which statistically relate a system's external and internal states. There must be some regularity in the environment for a system to be able to compare the model with reality and act in order to reduce differences. See Friston's (2010) free energy principle.

limited in what they are, what they can become, and their capacity to connect. Luhmann (46) states:

> The term "immanent limitation" refers to the internal complexity of the elements that are not available to the system, which at the same time makes possible their "capacity to unify." In this respect, complexity is a self-conditioning state of affairs: in that the elements must already be complexly constituted in order to be able to function as a unit for higher levels of system formation, their linking capacity is also limited, and thereby complexity reproduces itself as an inescapable given at every higher level of system formation.

As said above, complexity can only be reduced by complexity. According to Luhmann, this creates a "complexity gradient" (250) that "stabilizes" the system/environment difference.

1.5 How do we Recognize Systems? Function and Functional Analysis

If systems emerge as reductions of complexity, then complexity must be regarded as a "problem" rather than a cause. Complexity represents a problem to be solved by system formation. The reason why systems exist in the first place is to solve a problem. Systems are thus not self-sufficient substances resting in themselves simply lying around in the world, but functional entities whose ontological status is dynamic and processual. They do not simply exist; they operate; they do something. Systems exist because complexity has been reduced in a certain way and with respect to a certain form of problem-solving.

Systems theory is a functionalist theory, not a causal explanatory theory. Systems are always viewed as solutions to problems, not as states mechanistically caused by other preceding states or what can be termed external causes. For this reason, cybernetics and systems theory have always seen themselves as distinct from regular science with its strict exclusion of circular causality and teleology. When studying linear self-regulating machines and biological systems, it became apparent that strict causal determinism could not be the last word in scientific explanation. Complexity is neither a cause nor an effect of system formation but a problem that different systems try to solve by transforming greater complexity into lesser, manageable complexity. Functional analysis does not look for causes but "explains" a problem by discovering functionally equivalent solutions. For example, the functionalist view of a chair as a possible solution to the problem of sitting comfortably discovers not primarily how the particular composition of the parts mechanically enables sitting but looks toward the goal that is to be achieved by the chair and asks what functionally equivalent solutions to the same problem could be, e.g., stools, cushions, benches, etc. This aspect of systems theoretical thinking is particularly emphasized by Luhmann (1995:54):

> In this sense, the functional method is finally a comparative one, and introducing it into reality opens up what lies at hand for a sidelong glance at other possibilities. In the end, it ascertains relations among relations: it relates something to a viewpoint on a problem in order to be able to relate this to other problem solutions. Accordingly, "functional explanation" can be nothing other than the ascertainment (in general) and exclusion (in particular) of functional equivalents.

In systems theory, we are no longer within the framework of the causal-deterministic explanatory model of the natural sciences. And we are no longer within the world of the classical Western ontology of substance. The world does not consist of things but of functions. The main questions are no longer: From which causal processes and preceding conditions do certain phenomena arise? We are neither no longer primarily concerned with discovering which events lead to this or that change, nor do we consider a mathematical model that allows for prediction and control to be an adequate "explanation." We do not ask what external conditions cause system properties to arise. Instead, it is a question of what external or internal conditions allow or favor certain system states. We are not interested in how the whole can be analyzed into parts and explained from the parts. And finally, related to this on certain levels, we need no longer worry about how cognition corresponds to external reality. The "fit" or "adaptation" of the inside (system) to the outside (environment) is a matter of viability and not of truth. The entire epistemological quandary of how the subject can know the object falls aside. Whatever information the system might construct, as long as this enables it to continue its operations, it is "truth" for the system. These questions arise solely within the classical paradigm of natural science and the Western ontology of substance and subject. It is well-known and often said that systems theory has done away with efficient and material causes and replaced them with final and formal causes.[26]

According to the paradigm of systems theory, functional analysis does not start from mechanistic thinking but from analyzing dynamic systems as ongoing solutions to problems. This has the consequence, however, that on the most general or "metaphysical" level, it must be

[26] See Stadler/Kruse (1992:135-38) for a comparison of the systems theory paradigm with that of classical natural science. Stadler and Kruse list five assumptions of the classical model: 1) the equivalence assumption (*causa aequat effectum*), namely, "that in causal processes, to which all processes of nature are attributed, the effect is always equal to the cause, both quantitatively and qualitatively"; 2) the continuity assumption (*natura non facit saltus*), which "disregard[s] the possibility of erratic changes in natural processes." 3) the mechanistic presupposition, according to which "organismic processes [are] interpreted in terms of the interlocking of the parts of a machine"; 4) the elementaristic presupposition, according to which "a complex system [can be] analyzed by breaking it down into its elements and examining the functioning of these elements. The function of the whole system is then explained by the composition of the detailed processes in the elements"; and finally, 5) the realist notion, namely, "that the physical world surrounding us is actually as it looks to us." Against these assumptions of the classical paradigm, Stadler and Kruse set the following results of systems research: 1) "the analysis of complex dynamical systems shows that in nature very often bifurcations occur at unstable equilibrium points, where minimal causes can produce greatest effects"; 2) in many dynamical systems "phase transitions occur which generate entirely new order structures at a higher level of analysis"; 3) against the mechanistic assumption, systems theory knows that "the components of dynamical systems [interact] with themselves and, given initial and boundary conditions, produce autonomous states of order without being externally imposed or necessarily achieved by a mechanism"; against the elementaristic assumption, "elementary microprocesses do not run independently of each other, but cooperate with each other and generate new qualities on a macroscopic level, which cannot be explained by the individual elementary processes"; and finally 5) "the realist basic assumption...is opposed by a constructivist one. Cognitive systems do not absorb (semantic) information from their environment but generate it internally within the system itself."

assumed that complexity is somehow "perceived" as a problem; otherwise, "God" would not have created the world from the original chaos. In most creation myths, God or the gods are always trying to solve a problem. Why is there something rather than nothing? If mythical or theological talk is unwanted, we can talk about "self-organization" and "emergence" to explain the appearance of order from disorder.

1.6 What are Systems Made Of? Order, Organization, and Structure

Suppose complexity is reduced, and a system emerges by putting a smaller set of elements than originally given into an order, i.e., into certain relations to each other. In that case, this order can be considered the *organization* or structure of the system.[27] The organization of the system makes the system as a composite whole more – and at the same time less – than the sum of the parts. It is more because the system exhibits behavior that is not contained in any of the parts or the sum thereof, and it is less because the functions of the system do not correlate with the quantity of the parts. A system can have many parts but few functions. It is the system's organization and not any physical property, such as a shell, an

[27] The concept of "organized complexity" goes back to Bertalanffy (1956). In contrast to the classical mechanistic view of causality as a linear cause-effect relationship, Bertalanffy was one of the first to develop the concept of system as a term for dynamic wholes, such as organisms, which cannot be grasped within the framework of classical physics. See the discussion in Kneer/Nassehi (1993:21): "Organized complexity exists when individual phenomena are not simply logically coupled with each other in a linear fashion, but when there are interactions between them. If this is the case, only the description of these reciprocal networking conditions is able to convey a picture of the unity of the sum of those individual phenomena. Disorganized complexity can thus be described as a linear concatenation of individual phenomena: From A follows B, from B follows C, etc. Organized complexity, on the other hand, cannot be represented according to this simple model because it is conceivable that A and B are mutually the conditions of their possibility: So then neither A follows from B, nor vice versa, rather A and B are given to each other by their reciprocity, by their relation which cannot be represented linearly. It should have become clear that the essential object of systems theory is the organizational form of the complex interrelation between individual elements." It is out of the complex interrelations of elements that behaviors appear that can be called "emergent." Furthermore, it can be argued, as Levin (2022) has done, that living systems are organized not only by genes but by a morphogenetic form or pattern typical of a species.

enclosure, a membrane, skin, walls, boundary posts, etc., that demarcates the system from the environment and thus maintains the system/environment difference. Therefore, it is misleading to say that the system consists of thing-like conceived elements. The "elements" of a system are not things or substances lying around in the environment waiting to be taken up into a system. The system does not consist of things but of its organization, the specific processes of selection, relationing, and steering that are responsible for "constructing" the elements as system elements in the first place. What there is in the elements that are not the product of selection, relationing, and steering – as we will see in a moment – and what is not controlled by the organization of the system does not belong to the system but to the environment. We will explain the notion of organization further when discussing the idea of *code*.

1.7 What Do Systems Do? Operation and Process

The organization of the system determines how the system *operates*, i.e., what processes the system performs. The organization should not be understood as a static structure. A system is not only composed of elements in certain relations to one another but also of operations or processes that depend on how the elements interact with each other. Operations or processes are what the system does, and all systems do something.[28] Since processes are always system specific, we turn to a concrete example to illustrate the preceding definitions.

1.8 The Mechanistic Model – The System as Machine

To illustrate the basic concepts of systems theory that we have presented so far and also the concepts that follow, let us look at some examples.

---

[28] It is based on this general characteristic of systems that the many methods and models of the vast field called "system dynamics" has emerged, which attempts to identify the elements of a system and model, often mathematically or computationally, how the elements interact and thus to be able to predict and control system processes. See for an overview: Wikipedia https://en.wikipedia.org/wiki/System_dynamics#Overview.

The system, let us say, is an air conditioner. Although this is a mechanistic model, a machine designed and built by humans, it nevertheless provides the opportunity to present properties that apply to all systems, whether natural, biological, or social.[29]

For simplicity, let us assume that the air conditioner consists of only three components. There is a thermostat that registers the room temperature, a cooling system that produces cool air and blows it into the room, and finally a heater that produces heat. These three parts make up a system, not merely because they are distinct parts but because they are put together in a certain way and operate in certain ways. In fact, they could be far apart from one another. For example, the thermostat could on the wall in a room of the house, the cooling unit on the roof, and the heater in the basement. The parts make up a system because they are related to each other in specific ways that allow certain operations.

1.9 Code and Information

With this simple mechanistic model in mind, we now introduce other basic concepts of systems theory. The assignment of the parts of a system to each other, i.e., the particular kind of relations they enter into among themselves, constitutes the dynamic structure or organization of the system. These relations have a selecting effect. In other words, the orga-

---

[29] The following example is not presented by chance as a cybernetic system. The machine was the basis of cybernetics. Wiener spoke of cybernetics as the study of control in the machine and the organism. Ashby's influential book, *Design for a Brain* (1952), emphasized the continuity between organisms and machines, while today, terms such as "cognitive architecture" or "artificial life" clearly associate machines and intelligence as matters for engineering and design. According to Flechtner (1970:10), the model of the machine is equally applicable to other systems: "If we now try to give a label to cybernetics itself in anticipation, we can start from the concept of the system. Certainly, machines and organisms, human beings and their communities can also be regarded as systems, as organized wholes whose parts are interrelated in manifold ways. And which themselves, as a whole, are interwoven with other systems and generally with their environment in numerous relations. Abstractly spoken, they are structured systems, which behave in some way to other systems, thus are not rigid but dynamic systems. Thus, we arrive at the preliminary determination: cybernetics is the general, formal science of the structure, relations, and behavior of dynamic systems."

nization selects certain elements from all possible system components, relates them to one another in certain ways, and finally controls or steers the operations of the system. An organizing principle that simultaneously selects, relates, and steers is what we would like to call a *code*. Since every system consists of elements that stand to one another in certain relations by means of which the processes of the system are enabled and controlled, it may be said that the principle of organization or structure of every system is a code. Since the term code is usually associated with signs and symbols, we must explain how we are suggesting that the term be used within the scope of general systems theory.

Although the concept of code generally refers to rules, laws, and fixed procedures that can be followed to translate signs or symbols or program a computer, not all codes are related to signs. There are, in addition to semiotic codes, which, as we shall see below, play an important role in the organization of systems of meaning, genetic codes, which organize biological systems, and physical codes, which, as in the case of a machine, organize physical systems, for example, our air conditioner. The concept of code, then, does not imply the concept of sign, as some semioticians (Sebeok, 1991; Deely, 1990) assume. However, it does imply the concept of "information." Generally, a code, in the sense that we use it as a concept of general systems theory, is a rule for translating information into action. A code can be many things, but in some way, common to all various meanings is the idea that a code contains a set of instructions or rules that allow for the "translation" of certain things or processes into other things or processes. In the most general sense, the code, as the organizing principle of a system, "translates" information into functions and operations. The tendency to conceive of all system operations as "semiosis" and thus to speak of sign use and communication in physical and organic nature is based on the continuities between systems. It overlooks the discontinuities generated by the emergence of different levels of order. On the level of meaning, where semiotic coding is the principle of order, a sign is not a signal. A sign represents something "as" this or that. A sign must be understood. It is not merely a cause or a trigger of some event, as a signal can be. If a code performs differential relations of signs through syntactic, semantic, and pragmatic selections, relations, and controls, we are dealing with a specifically *semiotic* code. We are dealing with a semiotic code only if the system's operations are constitutive of meaning. When semiotic codes organize systems, they are *meaning systems* rather than physical or biological systems. This does not mean that "cognition," "problem-solving," or "modeling" cannot occur in biological systems. In fact, as we shall see, some of these concepts have their natural home on the biological level of emergent order and lead to problems when applied to the level of meaning. It should be noted that general systems theory commits meaning systems to "understand" everything as a system, including itself. Whatever kinds of systems and codes there may be other than semiotic coding is a question that is asked and answered only within semiotic coding.

In our example of the air conditioner, the code, i.e., the blueprint of the machine based on the laws of nature, relates certain parts made of certain materials and associates them with certain other parts in a certain way. This is selection and relationing. What about steering, the third necessary component of systemic order? What does the system do? How does the system operate? First, not only are certain parts, heating and cooling units, and thermostats selected and related, but also certain events in the environment are selected and assigned to specific values of the thermostat. Let us assume that there are only three values on the thermostat: "too hot," "too cold," "comfortable." The thermostat is built in such a way that out of all possible events in the environment, only these three make up the system-relevant environment and can interact with the system. Nothing else that happens in the environment, how many people are in the room, what they are talking about, whether music is playing, etc., is relevant or can become *information* for the system. The thermostat is one element in a system related to other system components, i.e., to the heating and cooling units, but it operates based on information.

The three values of the thermostat are translated or assigned to specific operations in the system, namely, turning on/off the cooling and heating systems or doing nothing. This is steering or control. As Wiener said, control via "communication," or information transfer, is the basis of cybernetics. The thermostat or whatever plays the same role in any machine, such as the flywheel of a steam engine, is called the "governor." The system solves the problem of an environment that has three degrees of complexity, too hot, too cold, and comfortable. It reduces complexity in terms of environmental events, elements integrated into the system, and relations among elements in such a way that certain operations are enabled, and by means of these operations, certain goals, and set points, can be attained. The system is self-regulating but not necessarily "adaptive."[30]

As a reduction of complexity, the code acts based on the system/environment difference and constructs the system and its specific environmental references. By considering system organization as code, the environment becomes structured, partitioned, and differentiated. This is to say, it becomes transformed into potential information. As mentioned above, there is still an absolute environment (environment-1), which is the entire world in which the house containing the air conditioner stands. However, it is the *system-relevant* environment (environment-2), which is the three values that the thermostat can register.[31] The code selects certain environmental events as potential information and constructs a system-specific environment, i.e., the environment for the system. Everything else that happens in the environment of the air conditioner, except for the three temperature events, is not "perceived" by the system as information but as noise. The noise in the room, whether people are present or not, whether it is winter or summer outside, whether the room is a schoolroom, a living room, or a government building; all this and infinitely(!) more is irrelevant to the air conditioning system and cannot become information for it or, what amounts to the same, cannot be "observed" by it. It can, of course, affect the air condition and disturb it.

---

[30] "Complex adaptive systems" are usually understood from the biological model of operational and informational closure, autopoiesis, reproduction, learning, and evolution. These are not usually characteristics of machines.

[31] Speaking of the environment as "relevant" and not merely "relative" to the system emphasizes that the environment does not consist of things or events but of information or noise.

What does it mean to say that only certain events in the environment become information for the system? The system/environment difference is to be thought of as a code in that only three environmental events have informational value, i.e., they are *differences that make differences* in the system. Information, according to Bateson (1979:99): "is a difference that makes a difference." This is illustrated by the fact that the information "too hot" does not "mean" or represent something intrinsically present in the world, but it refers to the difference between values on the scale of the thermostat. Information is not a thing but a relation, or as Bateson puts it, a difference, and differences are not things:

> The difference, being of the nature of the relationship, is not located in time or in space. We say that the white spot is "there," "in the middle of the blackboard," but the difference between the spot and the blackboard is not "there." It is not in the spot; it is not in the blackboard; it is not in the space between the board and the chalk. I could perhaps lift the chalk off the board and send it to Australia, but the difference would not be destroyed or even shifted because the difference does not have location. (Bateson 1979:98)

There are, of course, innumerable differences on a temperature scale, but only those are information that are selected as relevant to the system by its constitutive code. Information, then, does not exist in and of itself. Information does not lie around in the world until some system stumbles upon it. Information is constructed by the system. The code constructs information by constructing differences which then change the state of the system and select operations. Again: for the system, the world does not consist of things but of differences, which the system constructs. This fundamental principle of all systems is the basis of what has come to be known as constructivism and 2nd order cybernetics, about which more will be said later.

Another important property of the code, apart from constructing the relevant environment, is constructing the system-specific elements. The code draws the system/environment boundary in such a way that the elements are not, as it were, first found lying around in the environment outside the system and then somehow gathered up and brought into the system. Coolers and heaters do not exist in the environment; they are constructed by the system with regard to solving a specific problem. The system itself constructs the elements of which the system consists. This means that the concept of elements, components, parts, etc., must also be understood in a system-relative way. There is not a set of existing elements, say atoms, pieces of wood, stones, etc. which are first discovered in the environment and then somehow taken up and made into building blocks of different systems. The elements become only elements *of* the system *by* the system. This is why "translation" can be used in this context and why the principle of system organization can be called a code. It also implies that there are different elements at different system levels; atoms, cells, organs, and signs are all to be considered as elements of certain systems on system levels, each coded – and that means constructed – in their specific ways. Organisms are made up of different elements than machines, and semiotic systems are, in turn, made up of different elements than organisms. A system of meaning does not consist of cells, tissue, skin, organs, neurons, metabolic processes, etc.[32] On each level of emergent order, the code "translates" elements into functions in different ways. This means elements, like relations, are not optically or substantively given but are constructed by the system through its specific code. We thus are left with a completely relativistic and functionalist notion of element:

> Then an element would be what functions for a system as a unity that cannot be further dissolved (even if, viewed microscopically, it is a highly complex compound). When one says "cannot be further dissolved," this also means it can constitute and change itself only by interrelating its elements, not by dissolving and reorganizing them. ... Elements are elements only for the system that employs them as a unit, and they are such only through this system. (Luhmann 1995:22)[33]

[32] It is a commonplace error to think that neurons in the brain construct meaning or that signs are elements of the brain along with neurons.

[33] This means that systems analysis has to pay attention to the level of emergent

Let us take another example to illustrate this general principle of systems theory. Consider, for example, a table as a system. The table, let us say, consists of a top and legs. The top and the legs are the necessary elements of the table system. However, the function alone determines what can serve as a tabletop and what as legs. Table tops can be made of any material that is solid enough to support weight, and the legs can be made of wood, steel, glass, plastic, etc., so long as they function to support the top. Even a person can serve as a table if they go on all fours and hold that position long enough for someone to use their back as a table. The elements of the table system are not just lying around in the environment until the code of the table stumbles over them and integrates them into itself. Instead, the code selects certain elements that can serve the functions that need to be fulfilled to create a working space at mid-body height.

The same is true of living organisms. Let us say the elements of an animal are the organs, the skin, tissue, the bones, etc. The organism does not find these things just lying about somewhere in the environment and then integrates them into itself, but it constructs them according to the genetic and morphogenetic code that organizes it. And finally, a language does not consist of signs that can be found in the environment, but out of all the possible sounds that the human voice can utter, only certain are selected to bear meaning. The same, of course, is true of non-verbal signs. Therefore, along with the principle that all systems are constructed by processes of selection, relationing, and steering that create a system/environment difference, we can say that it is a second principle of general systems theory that *all systems construct their own*

---

order on which it operates. If one disregards level differences, e.g., by analyzing communicative actions neurophysiologically, then one abolishes the system/environment difference, i.e., the code that is constitutive for the system, and subordinates the system to an alien code, where one then can no longer compare it with the system's own elements, e.g., signs, but with elements foreign to the system, e.g., nerve impulses. For this reason, it has become commonplace to speak of a "hard problem" of consciousness. One cannot find meaning and experience in the brain because meaning is a higher level of emergent order than life. Meaning constructs meaning, and life constructs cells, organs, metabolic processes, etc. The entire materialism/idealism controversy that has plagued Western philosophy for ages and dominates cognitive science until today could be avoided by carefully distinguishing levels of emergent order.

elements by means of a code that "translates" elements into functions and operations.[34]

1.10 Cybernetics and Feedback Loops

The operations of a system are outputs that have effects on the environment. For example, our air conditioner turns on the cooling system or the heating system, which changes the temperature in the room so that "comfortable" again registers on the thermostat. From the system's point of view, it does not "know" what is happening out there in the environment. It only "knows" that the values on the thermostat are changing. According to the code, which is its ordering principle, the system operates until the value "comfortable" is registered again, i.e., until new information triggers new processes.

This form of operation is called a feedback loop, a control loop, or circular causality. The output of the system into the environment is received back into the system as input. Outputs become inputs. This is the loop. It is feedback because it "informs" the system of the effects of its outputs, allowing the system to control its operations such that the goals or setpoints it has can be attained. Thus, we can speak of a cybernetic or self-regulating system. Although all systems are self-regulating, the kind of self-regulation we are talking about depends on the code constituting the system because output and input are linked according to the possibilities of the code. The system could cause many things in the environment to happen through its operations. The air conditioner could produce many things, such as noise, stench, high costs, etc., but because of the code constituting it, these are all neither relevant nor "perceptible" to the system, i.e., they have no informa-

[34] This does not imply that there are no constraints on what elements can be translated into what functions. The laws of nature and characteristics of matter and energy are constraints for living systems, just as the laws governing living systems constrain meaning. Despite constraints on the part of the substrate, higher-level codes contextualize and constrain systems on lower levels. Life, for example, can do things with matter that matter could not do on its own, and meaning can do things with matter and life that neither could do on their own. A case in point is artificial intelligence and, of course, all technologies.

tional value; they do not become inputs or feedback, unless, of course, they are so selected.

1.11 Purpose, Teleonomy, and Homeostasis

Because the system has a goal, setpoint, and some information that when it is registered does not trigger an operation, i.e., it does nothing; it looks to an observer as if all the system's operations are aimed at achieving this one state over and over again. The system aims at stability or equilibrium. As long as nothing triggers some operation, the system does nothing. This value is then called the setpoint or target value because the system appears to act as if this value is to be maintained. Whenever relevant values are registered, e.g., "too hot" or "too cold," the system operates until it registers its target value "comfortable" again. For this reason, it can be said that cybernetic systems are goal-directed or teleonomic systems. The target value guides (controls!) the operations of the system and thus appears as a "purpose," a "goal." Moreover, because system operations are undertaken only to bring the system back to the same point, i.e., to (maintain) a stable equilibrium, such systems are called homeostatic systems or systems that operate to maintain a specific constant state.

This is not only true for cybernetic machines like the air conditioner. Organisms are the prime example of homeostatic systems. They act to maintain their equilibrium or, as we will say later, their "autopoiesis," for example, a certain oxygen level in the blood or a certain body temperature. In the case of living systems, the set points are not set by humans according to a mechanical code, but they are set by the genetic code and by interactions with the environment. However, it must be noted that organic systems never do nothing. They are constantly operating to ensure their further operations. This is called "autopoiesis," which we will discuss later. Furthermore, living systems strive for stability and maintenance of certain parameters, but they also generate instability and contingency through mutations and interactions with the environment. Our air conditioner does not do this. There is something about life that does not strive for stability, order, regularity, and predictability. This is, as we shall see, also true of meaning systems. To the extent that human beings and societies may also appear to maintain stability, they can be modeled as homeostatic systems, that is, systems that operate to maintain given parameters.[35] According to the theory of social systems that Luhmann proposes, social subsystems such as economics, politics, jurisprudence, education, business, etc. are guided by autopoietic, operational, and informational closure. The question of whether the respective codes of social systems operate homeostatically is a matter of interpretation. Do human societies strive for stability or change? Can history be understood as social and cultural evolution? We will return to these questions later.

1.12 Functional Differentiation, Subsystems, and Internal Complexity

Up to now, we have mainly dealt with the relationship between the system and the environment. If we now look at the system alone without reference to the environment, it becomes apparent that it is *functionally differentiated*. The system has different elements that perform various operations with different functions. In the case of our air conditioner, one function is to cool, the other is to heat, and the third is to register the temperature. For these three functions, the system has different *subsystems*, namely, the cooler, the heater, and the thermostat. The system, therefore, is not simple. It has an internal complexity that has come about through a process of *differentiation*. According to Luhmann (1995:53), differentiation is "the establishment of new system/environment differences within the original system."

The cooling system, the heating system, and the thermostat can be said to be *subsystems* within the overall system of the air conditioner. Similarly, the organs and functions within a living system that are essential to its viability can be seen as arising from the differentiation of

35 The mathematical model of such systems is what the "free energy principle" proposes. See Friston (2010).

subsystems.[36] The idea of a subsystem implies not only that it is part of a larger system but that it is a system itself and not merely an element. We would not say that a table's top and legs are subsystems, but a cooler and a heater are. Nonetheless, the legs and top are distinct functions and, from this perspective, constitute the internal complexity of the table. To the extent that internal complexity depends on the possible states of the system, and these, in turn, depend on the functional subsystems it has, internal complexity is not merely a matter of the number of elements in a system.[37] Furthermore, as in the case of the table, it is often unclear to what extent a subsystem constitutes a function within a system or whether functional elements that do not seem to fulfill all the requirements of a system still contribute to the internal complexity of a system. Finally, following Ashby's law that only complexity can reduce complexity and given the tendency of systems to attempt to reduce complexity, there looms on the horizon of systems theory the problem that, at a certain point, systems will become too complex to maintain their coherence and disintegrate into the environment or another possibility is that a higher level of order will emerge.

1.13 Contingency and Degrees of Freedom

Because the system has internal complexity, it has different possibilities for operation or action. In the case of our air conditioner, it can either cool or heat. This means that in a certain sense, the operation of the system is not determined, i.e., it does not have only one operation. If it can have different operations, there is *contingency*. Contingency means "it could be different." For example, the system could act this way or that way. If we don't know what the thermostat is currently registering, and if we don't know how the system is constructed – that is, if it were a "black box" – then we don't know what the system will do. It could

---

[36] Miller/Miller (1990) identifies twenty different subsystems necessary for life.

[37] Klaus and Liebscher *Wörterbuch der Kybernetik* define complexity as follows: "The complexity K of a system is directly proportional to the number n of its elements, the number z of the possible states of these elements, and the number k of couplings between the elements"(1979:314).

be heating or cooling. The term contingency denotes this possibility or indeterminacy. Contingency means that the system is not mechanistically determined such that only one course of action, a "reaction," is necessary. If we have a system whose operations are not causally determined, as they are for our air conditioner, these actions can be said to be "free." If there are two options, one can say the system has two *degrees of freedom*.[38] The more internal complexity a system has, the more degrees of freedom it has. And since there is a direct correlation between information and operation, the more information a system can construct, the more degrees of freedom it will have. Only systems that have internal complexity can exhibit contingency and can be said to have degrees of freedom. In the case of our air conditioner, we would have to assume that it could react to a drop in temperature in different ways and that the "decision" of which way to solve the problem is left to it and not the environment.[39]

Contingency is, in many respects, just another word for complexity. Contingency, for example, also represents a problem for the system; the problem of "deciding" what it should do. This amounts to reducing contingency, just like the emergence of the system in the first place can be thought of as reducing complexity. When the system "chooses" which course of action it will take, it has reduced contingency. A decision can be defined as uncertainty reduction. Out of the different possibilities, only one is selected. Before the decision one didn't know what the system would do. Afterwards, the situation is no longer uncertain. We can distinguish between system contingency and environmental contingency. The environment consists of a multitude of possible events that could happen. Taking our air conditioner again as an example, it could get too hot or too cold, or it could stay comfortable, and of course, many other events that are irrelevant to the system could happen.

---

[38] Mechanical or physical systems of course also have degrees of freedom, but the term refers to the number of variables or independent parameters which determine states of the system. See Wikipedia for an overview https://en.wikipedia.org/wiki/Degrees_of_freedom_(mechanics).

[39] Levin (2022) defines "intelligence" as the ability to solve a problem in more than one way and finds this ability throughout the living world from the single cell to human brains. For Levin life as such is intelligent.

Depending on what happens in the environment, the system reacts. But the system does not "know" in advance what will happen. From the system's point of view, the behavior of the environment is undetermined and, in this sense, uncertain and contingent. Some systems have no internal complexity and, consequently, no contingency. They are completely determined and "simple."[40] The environment, however, is always complex and contingent, and the primal substrate we postulated at the beginning of the world is at once absolutely complex and absolutely contingent.

1.14 Indeterminate and Determinate Complexity/ Contingency

As noted above, there are three kinds of complexity and contingency. Two of them concern the environment: 1) absolute or (world) complexity or contingency, i.e., anything at all can happen, everything is equally probable, and there is no reason why there is something instead of nothing, and 2) complexity/contingency of the system's environment, i.e., not everything can happen, and only certain events that do regularly happen are considered relevant for the system. Let us call these two forms of complexity indeterminate complexity/contingency and determinate complexity/contingency. For the air conditioner, only three values exist as possible or perceptible events in the environment: too hot, too cool, and comfortable. These three states represent the determinate complexity/contingency of the environment. Determinate complexity forms the *horizon* of possible events, the "perceptual field" of the system, or everything that can become information for the system. But there are other events in the environment than just these three. For example, an earthquake could destroy the building, or it could burn down, or it could be deliberately demolished, or it could be noisy in the room. Many other things could happen, and all of these would not "mean" anything to the air conditioner because its constitutive code has not determined such possibilities to be possibilities for it. The

[40] Heinz von Foerster (1992) called such simple systems "trivial machines."

code makes certain events in the environment relevant to the system and leaves everything else undetermined. It constructs a system, i.e., creates a system/environment difference, by reducing the indeterminate complexity of the world to determinate complexity/contingency of the world for the system. Every system has its environment, and no system faces the task of selection, relationing, and steering over against absolutely indeterminate complexity. Determinate complexity constitutes the constraints that condition system organization. The universe is always already ordered in some way. And so, it makes sense to speak of "natural selection."

It should be kept in mind, that determinate complexity/contingency depends directly on the internal complexity of the system. The code, as said, selects only certain events from all possible events as relevant to the system. The system thus constructs its environment, which consists of the possibilities that the code of the system has determined to be possibilities for it. This is not to say that the system can wish for any environment. The environment contains many, perhaps infinite, constraints that are not at any time all relevant for the autopoiesis of the system. These are ignored, or if relevant constraints are ignored, the system disappears. For all such events, the system is "blind" because it has no differentiated internal subsystems or operations corresponding to them. If a system has no internal complexity, its environment can also have no determinate complexity. In such a case, there is only one relevant information and only one operation corresponding to it. One cannot even speak of "information" in this case but "cause" or "trigger." Thus, the determinate complexity/contingency of the system's environment is related to its internal complexity/contingency, i.e., to its functional differentiation. The more internal system complexity, the more determinate environment complexity. The more complex a system is, the bigger its world. We will return to these ideas when discussing the conditions under which life and meaning can emerge.

1.15 Autonomy, Learning, Emergence, Levels of Emergent Order, Evolution, Self-Organization

Imagine a much more complex system than our simple air conditioner. Suppose the system has not only the two subsystems with operations of cooling and heating available in its repertoire but also other possibilities, e.g., instead of turning on the cooler when it is too hot and the heater when it gets too cold, the system could be set up to open and close the windows depending on the outside temperature. Or one could imagine that an alarm would go off when it gets too hot, alerting people of the danger or automatically contacting the fire department. All of these and other conceivable possibilities are functional equivalents that become possible because of the functional differentiation in the system, i.e., greater system complexity.

Let us further assume that the system can "decide" which functionally equivalent operations it wants to perform. Depending on the situation in the environment and on the internal state of the system, it could "choose" this or that operation. One time to open the windows, another time to send people out, a third time to turn on the cooling, and so on. The system has sufficient *autonomy* to solve a problem in different ways.[41] Let us further assume that the system is capable of *learning*, based on experience, which of these functionally equivalent operations works best in each case. We could also assume that the system is capable of a certain *planning*. It can predict probable environmental states and adjust its operations to meet future needs. Furthermore, we may assume that the system is *self-productive* and can maintain itself and produce its parts, e.g., by being connected to an automated factory that produces spare parts and other similar systems according to instructions from the system. And finally, let us assume that the system can even *reorganize* itself such that it can still function when the environment changes in such a way that its former organization no longer functions; that is, it can *adapt* to environmental changes by changing its organization and

41 As mentioned above, Michael Levin (2022), following Maturana and Varela and William James, calls the ability to solve a problem in more than one way "intelligence" and equates life with cognition or intelligence.

becoming something different than what it was before.

With these assumptions, we notice that the model of a machine and the basic concepts that apply to it and can be derived from it are no longer adequate to describe the system. Despite all the continuities and similarities that cybernetics has discovered between self-regulating machines and organisms, there are still fundamental discontinuities and differences between mechanisms and life. When a system has many different behavioral possibilities, the ability to learn from its actions, and the ability to reproduce itself, anticipate the future, and plan, it is no longer what is usually understood to be a machine but a living being.[42] At a certain point, the system's increasing complexity forces us to speak of an entirely different form of systemic order. The system now becomes a living organism. There is a need for a specifically *biological model* instead of a mechanistic model. This doesn't change, even if we can also speak of "artificial" life and "artificial" intelligence in the case of engineered systems exhibiting the characteristics we have just described but which have been constructed by humans and are products of "design." Such artificially living and intelligent systems are modeled on living systems and designed to operate similarly to living systems. At what point do artificial systems become "life," or are we forced to redefine what "life" means, or must we create new hybrid categories to account for beings that heretofore did not exist or that always existed but were ignored?[43]

---

[42] Von Foerster's (1992) well-known distinction between trivial and non-trivial machines could be cited here. The trivial machine is a determined input-output loop, whereas the non-trivial machine is not determined by inputs but has autonomy with regard to outputs and therefore is unpredictable. The non-trivial machine is non-linear and historical. Von Foerster and cybernetics in general has chosen to emphasize the continuities between machine and organism. Michael Levin (2022) also emphasizes continuity and proposes that there is agency and intelligence down to the atomic and even subatomic levels as well as unbroken continuity between machine and organism. This can be admitted as an effect of scale of observation while also affirming that emergence, major transformations, or discontinuities are apparent at larger scales.

[43] See for a discussion Deplazes/Huppenbaur (2009); Wikipedia *Artificial Life* https://en.wikipedia.org/wiki/Artificial_life. Here to the work of Michael Levin (2022) should be included as challenging the clear distinction between mechanism, life, and mind, and of course, Latour has challenged the traditional distinction between the artificial and the real, things that are constructed by humans and things that construct themselves.

To introduce and define further basic concepts of systems theory, we now have to jump to a higher level of emergent order and speak of living systems. Our models necessarily become biological. This jump is not arbitrary but results from the development of the complexity of our subject. As we saw, the system we were describing became increasingly complex, and suddenly it reached a point where it "jumped" to a higher level of order. This jump can be called *emergence*.[44] Since living systems are not an effect of physical causes, they can be said to be *self-organizing*.[45] There "emerges" something that cannot be derived from the elements and operations of systems on the merely physical or mechanical level of order. Even today, biologists do not know what life is or how it originated. What is known, however, is that the emergence of life has something to do with complexity.[46] It would seem that when a certain degree of complexity arises such that existing codes and forms of organization can no longer reduce it, a jump occurs to a higher level of emergent order and a more complex form of coding.

The result of such a fundamental transformation can be called the transition to a *higher level of emergent order*. This occurs on the basis of a new kind of coding. DNA is the code of living systems.[47] Once life has emerged as a level of order beyond physics and chemistry, many new aspects of systemic order appear. Living systems are autopoietic, self-referential, informationally and operationally closed systems.[48] They operate to maintain their operations, their "autopoiesis," that is, self-production. They do this within an environment that "selects"

---

44 Heylighen (1989:382) defines emergence as "a qualitative change, where a new organization or system appears, with properties (potential appearances) that did not exist in the old system."

45 Self-organization is a widely documented phenomenon found in many kinds of systems on all levels of emergent order. For an overview, see Wikipedia: https://en.wikipedia.org/wiki/Self-organization.

46 See, for example, Assembly Theory by Sara Walker and Lee Cronin. https://en.wikipedia.org/wiki/Assembly_theory

47 The work of Michael Levin (https://www.drmichaellevin.org/) has raised the possibility that electronic morphogenetic coding or bioelectric networks also play an important role in organizing living systems. Levin uses the metaphor of hardware and software to explain the relation of DNA to morphogenetic electrical patterns.

48 Maturana/Varela (1980; 1987).

for viability, that is, successful adaptation. Since the environment is constantly changing and is changeable, contingency comes from the outside. But the genetic code is subject to more or less random *variation* that changes the organization of the system such that unforeseen possibilities of action internal contingency can lead to successful adaptations. This kind of interaction and relation between system and environment can be termed *evolution*. Evolution means that the autopoietic, system-building reduction of complexity allows for variability that is then selected by the environment. Evolution implies that the system reduces internal and external contingency so that autopoiesis can continue. The environment allows some mutations of the organism to survive and reproduce. This is called natural selection. Variation and natural selection are principles that all genetic coding is subject to. They determine the relationship of the living system to its environment. It should be emphasized that evolution is not emergence. Evolution is the ongoing play of variation and selection, allowing the genetic and morphogenetic code to explore the possibility space of any environment. However, the emergence of life as such is not a product of evolution. Instead, life is what makes evolution possible. Only living systems evolve. Life did not evolve from physical or chemical systems. Life emerged from physical and chemical complexity as a higher level of order and a fundamentally different form of coding.

Generally, organic systems have emerged from inorganic systems, just like meaning systems have emerged from life. The genetic code organizes systems that are subject to the pressures of variation and selection. The semiotic code organizes meaning systems that are not subject to evolutionary pressures and should not be modeled as living systems. Meaning is not subject to variation and selection and cannot be said to "adapt" to any environment. For this reason, as we shall see, it is misleading to speak of cultural or social "evolution" or social Darwinism. It is, however, commonplace to do precisely this. Despite all tendencies to the opposite, concepts such as evolution and adaptation have no place in social theory. Considering that everyone, at least since Spencer, assumes there is such a thing as social evolution and that evolutionary theory applies to society and culture, to claim the opposite as we do, is so unusual that we will have to argue for these claims at length later in this book. We will discuss semiotic systems that represent a higher level of emergent order, and which cannot be understood on the basis of a biological model. For the moment, it is sufficient to note that the dynamic relations of the system to the environment are different on each level of emergent order, even though there are continuities and analogies that permit using certain common concepts. The claim of general systems theory, which we are critically testing in this book, is that the basic systems theoretical concepts on one level can be carried over to the next higher level. Just as one can speak of self-organization on the level of physical systems, this concept can be applied to living systems and even meaning systems. Systems theory offers general concepts like self-organization and emergence that can be fruitfully applied to understand various phenomena. We will see that this may be the case for certain systems theoretical concepts, whereas for others, the systems model breaks down.

With regard to emergence and self-organization, let us look at some current definitions. Ebeling (1989:17) writes:

> By self-organization, we mean an irreversible process that leads to more complex structures of the overall system through the cooperative action of subsystems. Self-organization is the elementary process of evolution, understood as an unlimited sequence of self-organization processes. In this sense, the processes on Earth and in the cosmos are generally evolutionary processes that can only be understood in the context of their history, i.e., the entire chain of causative self-organization processes.

Closely linked to self-organization is the concept of emergence. Emergence is defined by Krohn/Küppers (1992:389) as follows:

> In the 'classical' sense, emergence means the emergence of new layers of being (life vis-à-vis inanimate nature or mind vis-à-vis life) that are in no way derivable, explicable, or predictable from the properties of an underlying level. Hence, they are perceived

> as 'unexpected', 'surprising', etc. In a modern version, we speak of emergence when a new quality arises through microscopic interaction at a macroscopic level, which is not derivable (causally explainable, formally inferable) from the properties of the components, but which nevertheless exists solely in the interaction of the components.

To a certain extent, emergence can be seen as a process of self-organization. This is because 1) systems show a tendency to increase internal complexity and 2) from the growing internal complexity of the system, which conditions increased environmental complexity, the need arises for a new organizing principle capable of solving the complexity problem. Only the system itself can solve this problem. The very idea of emergence implies self-organization since emergent phenomena are not caused by external factors. At a certain point, chemical or molecular systems became so complex that this complexity could only be reduced by a new kind of organizing code, the genetic code. The genetic code is more complex and variable than the physical code, so it can do things with matter that matter alone could not do. Life organizes matter in ways that matter would not be able to do by means of physical codes alone. As we shall see, the semiotic coding of meaning systems is more complex and variable than genetic coding and can therefore do things with life that life alone could not do. This explains how technology and genetic engineering are possible. It also explains why meaning systems can produce artificial intelligence on a non-biological substrate. But more of this later. In a certain sense, it could be said that life is more "powerful" than matter and meaning more "powerful" than life.[49] But as we shall see, these different forms of order are "powerful" in different ways. They do not all do the same things in the same ways.

[49] For meaning systems, life is cheap. Following Socrates, it can be said that the life without meaning is not worth living, and history demonstrates how many offer their lives for their beliefs.

1.16 Excursus: The Self-Organizing Universe

As a scientific paradigm and a supposedly universal theory, systems theory is also a worldview. Its claim to universality has a visionary character. This can be seen in the following story of the self-organizing universe.[50] The concept of self-organization is comprehensive and has been used to describe everything from nano-structures up to the cosmos. With regard to systems theory, self-organization always refers to a system of some kind, that is, a composite of parts related in specific ways such that the composite can be distinguished from an environment and specific tasks can be accomplished. The relations among the parts are the organization of the system. The concept of self-organization implies that this organizing principle and process comes from the system and not from any external cause. God cannot create a system. The system is the agent pulling itself up, as it were, by its bootstraps. In self-organization, there is always some spontaneity, autonomy, autocatalytic event, etc. The concept does not claim, however, that nothing outside the system "triggers," "occasions," "influences, or "constrains" self-organization. It does not claim that something comes from nothing. The trigger or occasion for self-organization is usually some change in the environment that often acts as a selector for new structures of the system, but it can also be some "mutation" in the system itself. In our story of the self-organizing universe, we are primarily concerned with the internal complexity of the system – which, of course, is always related to external complexity – as the driving force behind the self-organizing transformation of the system.

As we have said, systems arise to solve the problem of complexity. What is most complex is what we have referred to as the primal substrate or medium, which is absolutely complex. In the primal substrate, all events are equally probable. This is the chaos before the creation of the world, that is before any forms of systemic order come into being. The first systems at the beginning of the world were purely physical. The principle of their organization was the laws of nature that structure physical

[50] See Jantsch (1980) for a detailed discussion of self-organization and evolution.

and chemical systems. Classical natural science did not acknowledge the possibility of self-agency of any kind. Things were passive until acted upon by external forces. The world did not consist of systems but of objects, things, and inanimate matter in motion. This changed with the systems paradigm. Systems theory subscribes to circular and teleological causality and holism and applies these explanatory models also in the physical sciences.

Generally, even for physical systems, it could be supposed that to reduce environmental complexity, the system must build up internal complexity. In cybernetics, the need to build internal complexity to respond to external complexity is called Ashby's Law.[51] To cope with environmental complexity, the system must create internal complexity through structural differentiation, that is, via self-organization. The more internal complexity a system has, the more environmental complexity it can successfully reduce. As in the air conditioner model, the system could successfully respond to a few different environmental events thanks to its internal differentiation into a cooler and heater. A living system can respond to many more external perturbations since it has greater internal complexity. Systems theory postulates an innate agency or tendency of systems to increase internal complexity, which lays the foundation for self-organization and the emergence of ever-higher levels of order.

Returning to our story, at a certain point in time, chemical systems became so complex that they could no longer be efficiently organized by the physical code. As a solution to the problem of over-complexity, life emerged based on genetic coding. Emergence is a radical form of self-organization that no longer merely changes structures and adapts to an environment; this is normal evolution. Emergence changes the principles that organize the system. Life is a fundamentally different kind of system than physical systems. No amount of physics is going to "explain" life. The methods, instruments, and theories successfully

[51] Usually stated as "only variety can destroy variety," Ashby also stated it as follows: "The larger the variety of actions available to a control system, the larger the variety of perturbations it is able to compensate." Ashby (1956).

used to observe and explain physical systems cannot describe living systems. Quantum mechanics will not explain life or consciousness. Not respecting this discontinuity in levels of emergent order leads to physicalist reductionism and, on the other side, to panpsychism.[52]

At a certain point, living organisms became so complex, for example, with the development of a central nervous system consisting of billions of neurons and trillions of connections, that only a new form of coding could reduce this complexity. Meaning or semiotic coding emerged. Meaning systems are an entirely different form of system than living systems. Nonetheless, it is a thesis of this book that systems theory, with few exceptions such as Luhmann, never clearly distinguished between life and meaning. Indeed, the idea that life is cognition, that cognition is a process of the brain, and that mind and nature are one has become a common motif in systems theoretical discourse and cognitive science. We will return to this problem later. For the moment, let us continue our story of the self-organizing universe.

The general tendency of systemic development dictates that meaning systems, that is, societies and cultures, become more and more complex by processes of internal differentiation. History tells us how originally simple societies, languages, forms of government, technologies, beliefs, etc., have become more complex and differentiated over time. Civilization is a story of progressive expansion of knowledge, art, technology, etc., which leads to ever more complexity that the system must somehow control. Presently, some voices claim that society has become so complex that it can no longer be organized by means of semiotic coding, and some new form of order must arise to prevent society from collapsing into chaos.[53] Typical symptoms of this precarious situation are the many global problems of climate change, migration, uncon-

[52] The seemingly endless debate between materialism and idealism can be traced back to this disregard for levels of analysis. For a critique of physicalism see the work of Bernardo Kastrup (https://www.bernardokastrup.com/). Levin's (see Fields et al 2022) panpsychism is anchored in application of the free energy principle to quantum systems.

[53] For an overview of the discussion see Wikipedia Complex Society (https://en.wikipedia.org/wiki/Complex_society)

trollable financial systems, pandemics, international conflict, etc. The rise of Earth Systems Science, ecology, and global governance can be seen as the attempt to model the world as a systemic whole that can, in principle, be predicted and controlled. Whether world complexity can be successfully analyzed by any system dynamics models and thus become accessible to prediction and control is an open question.[54] Not only is it an open question, but it is also a controversial issue because increased control correlates with decreased freedom. A world ecological system that is in every possible way predictable and controllable seems to resemble the most radical form of totalitarianism.

If it is not possible to mathematically and computationally model the world, one may suppose that the problem of societal over-complexity will be solved by a fourth leap to an even higher level of emergent order, a level of order higher than meaning. Of course, we cannot know what this would be since whatever we can know is known on the level of meaning. Complex molecules could not have foreseen life, and complex living structures could not have foreseen meaning. We are "trapped" in our semiotic order; we cannot think of any other ordering principle beyond meaning. Such a possibility is not within the horizon of our perception. As Wittgenstein said, the limits of one's language are the limits of one's world. We cannot imagine what the next developmental leap will look like. Every attempt to imagine a transhuman form of being is done within the realm of meaning. Otherwise, we couldn't talk about it. Whatever might appear to solve the problem of the over-complexity of meaning, whether it be some form of transhumanism, cyborgs, robots, or AIs, it appears for us as an open horizon of possibility, as the transcendent par excellence, as the unmastered contingency of human existence.

The moral of this story is that generally, the development of systemic order runs in the direction of ever higher system complexity and ever higher contingency, variability, and transformability. It seems that the goal is to absorb the complexity of the environment into an all-encompassing system that in every way equals environmental complexity.

---

[54] Wofram (A New Kind of Science, https://www.wolframscience.com/nks/

As Ashby predicted, what was outside is reproduced inside. And as Jantsch (1987:181) says, "Not in the construction of hierarchy, but in the unfolding of complexity lies the real work of evolution." Complexity is not only at the beginning of the world but is apparently also the final and apocalyptic goal of its development. Systems become increasingly complex to cope with the absolute complexity of the environment until they become so complex that they no longer differ from the environment, and the constituting difference from the environment disappears.

Suppose we grant that internal complexity is formed against the background and as a response to environmental complexity. In that case, we must ask if it will come so far that systems become equally complex as their environment, i.e., that they finally and definitively solve the problem of complexity. The system would be exactly as complex as the environment. Indeed, this is the stated goal of datafication or applying descriptive, predictive, preventive, and prescriptive analytics to big data and modeling reality by so-called "digital twins." Can this be done for small subsystems such as a machine, a factory, a city, the human body, etc., and finally for the Earth System? If so, there would be no difference between the system and the environment. System complexity would equal environment complexity. But if there were no difference between system and environment, there could be no system because systems are constituted necessarily by a difference to the environment. Is the goal of evolution its dissolution, the destruction of all systems, and the return to absolute entropy? Suppose evolution aims at returning to the state from which it started, namely the equal probability of all events. In that case, we have a story that is like physical theories of the "big bounce,"[55] namely, the universe comes into being, expands, and then collapses only to come again into being and so on infinitely. Accordingly, systems theory is open to evolutionary as well as devolutionary interpretations. We do not know whether we are going up, down, forward, or backward. The entire story can signify the fulfillment of all possibilities and a senseless cycle. There seems to be no reason internal to the theory for optimism. Finally, it is a matter of faith that we must leave to the

---

55 See the Wikipedia article: Big Bounce https://en.wikipedia.org/wiki/Big_Bounce.

"theologians" of systems theory.[56] Another option we will explore in this book is that when systemic order becomes too complex, the world does not collapse into chaos. Instead, systemic order is transformed into network order. At a certain level of complexity, systems become networks. We will argue that this is what we are currently witnessing in all areas of society.[57]

1.17 The Biological Model – The System as a Living Being. Autopoiesis, Allopoiesis, and Autonomy

Before taking off on our excursion into the implications of the idea of the self-organizing universe, we introduced a new kind of system that brings new basic concepts into general systems theory. This was the living system. According to Maturana (1985), a system that produces itself is called *autopoietic*. "Auto-poiesis" means "self-producing" from the Greek "*auto*" = "self" and "*poiein*" = "to make,", "to bring forth." Systems, as we saw, consist of elements in certain relations, which make certain operations possible. If the operations of a system consist in producing the elements and relations of which it consists – and do nothing else – then this system is an autopoietic system. It is a system whose goal is nothing other than producing and reproducing itself. In a definition that has already become classic, Maturana says:

> An autopoietic machine is a machine organized (defined as a unity) as a network of processes of production (transformation and destruction) of components that produces the components which: (i) through their interactions and transformations contin-

---

[56] For an optimistic interpretation see e.g., Moltmann (1985:213), who equates the absolute world-complexity or the environment of the universe with God: "All individual matter and life systems as well as their communicative interrelationships as a whole ek-sist into a transcendence and subsist from this transcendence. If we call this transcendence of the world 'God', then we can say tentatively: The world is in its details and as a whole a God-open system. God is its extra-worldly environment, of which and in which it lives. God is its extra-worldly anteroom, into which it develops. God is the origin of the new possibilities from which it gains its realities."

[57] The doubtful reader can scroll forward to Part III of this book to see many illustrations of this tendency.

> uously regenerate and realize the network of processes (relations) that produced them; and (ii) constitute it (the machine) as a concrete unity in the space in which they (the components) exist by specifying the topological domain of its realization as such a network. (Maturana/Varela 1980:78)[58]

> Consider, for example, the case of a cell: a cell is a network of chemical reactions that molecules produced in such a way that they 1. produce, or recursively participate in, through their interactions, the very network of reactions that they themselves produced, and that 2. realize the cell as a material entity. The cell, therefore, remains as a natural entity, separable topographically and operationally from its environment, only so long as this organization is consistently realized by perpetual turnover of matter, regardless of the changes in its form or the specificity of the chemical reactions constituting it. (Varela et al., 1974:188)

What characterizes a living system is that its operations are not directed to any goal beyond itself. This is what "self-reference" for a living system means. A system whose output or product is something other than the system itself, such as the air conditioner, which outputs heat or cold and thus aims at producing a particular state in the environment, is not autopoietic but allopoietic. The goal of the non-living system is not the production of its own elements and organization. Nevertheless, there are similarities. Self-organizing systems operate like allopoietic systems in that they "try" to maintain their set point or at least some steady state or equilibrium. Thus, they can be said to be homeostatic and cybernetic, but unlike allopoietic machines, living systems are

[58] See the explanation by Roth (1987:264): "...the production of the components must take place in a more or less precisely defined way, and in such a way that the component A (partly together with other components) creates exactly those conditions under which component B can emerge, which in turn (together with others) creates exactly those conditions under which C can emerge, and so on, until finally a certain component (again together with others) creates the conditions under which A can emerge again and the whole (in itself many times intertwined) cycle of mutual production and repair can be run through again." This entire process has since been described in detail in terms of RNA protein synthesis, ribosomes, transcription, and translation.

oriented to their own organization, i.e., their set point is not imposed on them from the outside but is set by their organization. They do not "operate." They "act." It needs to be emphasized that they do not act to cause some effect in the environment, but instead, to be able to continue to act, to maintain their organization, or their autopoiesis. Life is agency for its own sake.

Whatever events in the environment autopoietic systems respond to, they respond not to change or to restore some state in the environment, but whatever they do, whatever the operations might be, the purpose is to enable the organism to continue its autopoiesis. This is important because one cannot model living systems as agents of stability and regularity. They do not seek the most probable outcome but the outcome most conducive to autopoiesis.[59] The air conditioner begins to operate when certain information is registered about the difference of the temperature to the setpoint. Such systems operate to minimize or eliminate differences between what their setpoint demands and what the environment is doing. Their agency is inherently conservative and stabilizing. On the contrary, the organism responds to inputs according to a genetic code that governs the autopoiesis of the system. For example, an organism can react to cold or heat by running away instead of trying to change the environment. It can respond by generating mutations that allow it to withstand greater or lesser temperatures. The living system is not concerned with the environment but with itself. Lions don't hunt zebras to reduce the population of zebras in the environment but to be able to go on hunting. Lions are only concerned with themselves. A system that is concerned with itself is autopoietic and also autonomous. Autonomy means that the system, by its own structure, determines the series of states through which it passes. The system's state changes are not controlled "from the outside," i.e., heteronomously or mechanistically, but by the internal genetic code.

---

[59] This is why Friston's free energy principle and its Bayesian mechanics do not offer a good model of living systems. Living systems are more than adaptive behavior.

1.18 Operational and Informational Closure, Self-Reference

The fact that the living system is concerned only with itself means that it is autonomous, implying that it is an *operationally closed* system. A system is operationally closed in a minimal sense if its states are determined only by its own structure.[60] Whatever the system does, it operates only by virtue of its own organization and, thus, in a minimal sense *of self-reference,* only in relation to itself.[61] Operational closure and self-reference mean the system's operations are directed toward itself. In this sense, the machine system of the air conditioner is also operationally closed and self-referential since it operates recursively according to its own structure. It is not concerned, and cannot be, by anything else that happens in the environment other than whether it is too hot, too cold, or comfortable. And in a certain sense, its goal is not to change the environment, even if this always happens, but to maintain its structurally given setpoints. In this sense, operational closure and self-reference trivially apply to all conceivable cybernetic systems since a system always operates by virtue of the organization/structure that defines it and which delimits the system from the environment.[62]

At the level of autopoietic systems, however, operational closure and self-reference imply *informational closure.* The system constructs information out of perturbations from the environment; there is no exchange of information between the system and the environment. An autopoietic

---

60 See Roth (1987:259): "Although materially and energetically open, it [the system AB/DK] itself determines its state sequences on the basis of its specific internal structure, which makes autopoiesis possible. ... Structurally determined and thus autonomous systems may be externally stimulated or 'perturbed,' but these influences do not determine the state sequences of the system." Roth (264-5) notes that the term "autopoiesis" has two meanings: self-production and self-preservation. The first meaning, self-production refers not only to living things but also to physical systems. Only self-preservation describes autopoiesis in living things.

61 The term "self-reference" can be applied to all recursive, cybernetic systems, e.g., also to the air conditioner. In this case it is of course not a question of self-consciousness. Self-consciousness is the form of self-reference that takes place at the level of meaning systems. Keeping this in mind would contribute to clearing up much confusion in debates between materialism and idealism (panpsychism).

62 Thus, Luhmann (1984:31) can understand his claim that "there are systems" as equivalent to the claim "there are self-referential systems."

system acts only to maintain its autopoiesis and thus refers only to itself in the ways in which its internal differentiation allows. Information is internally constructed out of undifferentiated perturbations coming from the environment. The eye of the frog constructs information out of quickly moving black smudges within its field of vision. The information says, "This is good to eat." The selection of a system-relevant environment, for example, the quickly moving black smudges, is done by the system and not the environment. According to Luhmann (1990b:321), "information...is not an input, but only the differentiation of connection possibilities" within the system amounts to information. Information is always only information for the system by the system. Information is generated internally in the system and not read from the environment. As noted, this need not mean that the system acts "self-referentially" in the sense of "self-aware," but only that it has no informational "input/output" interface. Environmental events have no *informational value* for an informationally closed system. They are merely "perturbations," i.e., undifferentiated disturbances that can only become information due to the system's organization or code. Of course, if there were no structure and regularity in the environment, the frog would never have been able to select quickly moving black smudges as information in the first place. Indeed, life would not be possible without regularities in the environment.

With regard to semiotically coded meaning systems, what has meaning and can have meaning must be coded by the system as meaningful. The meaning system does not find signs lying about in the environment and only needs to read them. If anything has meaning, and one can talk about it, this is because it has been coded as information. As Wittgenstein put it, "Whereof one cannot speak, thereof one must be silent." Operational and informational closure and self-reference are constitutive of all autopoietic systems. In the case of systems that do not produce themselves, that is, allopoietic systems, they are informationally open and, therefore, can "know" from the environment what its code has selected as relevant. The autopoietic informationally closed system does not "know" anything about the environment, and when it comes to organisms, instead of meaning systems, the system does not even "know" about itself. Maturana (1987:105) describes this for an organism and its nervous system as follows:

> With respect to their state dynamics, however, the organism and nervous system operate exclusively as closed systems that merely generate structure-specified states, much as a pilot behaves in instrument flight. When the pilot must fly and land without visibility, he must keep the indicators of his aircraft's instruments within well-specified limits or follow a series of specific variations. When the pilot leaves the aircraft after landing, his wife and friends might come up to him and say, 'That was a wonderful flight with an excellent landing!' We were afraid because of the fog!' The pilot might reply, 'What flight? What landing? I didn't fly, I just kept my display units constant in certain areas.' In fact, there was a flight only for an external observer, and exactly this happens with an organism and its nervous system.

When speaking of operationally and informationally closed systems, this does not mean they are closed to everything. As Bertalanffy pointed out, all systems are open with respect to the exchange of matter and energy. Allopoietic systems are open to informational steering from the environment and are not operationally and informationally closed. Autopoietic systems, above all living organisms, are operationally and informationally closed self-referential systems.

We can now add a third principle of general systems theory in addition to the first two. The first principle states that all systems are organized by selection, relationing, and steering, which constructs a system/ environment difference. The second states that all systems construct elements that are not things but functions. And now, third, all systems, at least those relevant for our purposes focusing on social systems, are self-organizing, autopoietic, self-referential, and thus operationally and informationally closed.

1.19 Structural Coupling, Adaptation, Viability, and Again Evolution

Structural plasticity, or the ability of an autopoietic system to change its structure even though it is structurally determined, is called *adaptation*. The term "adaptation" is misleading in that it presents the image of an intentional act aimed to optimize responses to the environment. However, an operationally and informationally closed system cannot "know" the environment *per se*. It can only react to impulses, disturbances, or perturbations coming from the environment according to its organization. If the system has the ability to change its structure or organization, what could be called "adaptability," and if indeed the structure of the system changes through interactions with the environment in such a way that the system can continue its autopoiesis, then this appears to an observer as if the system has "adapted" to its environment. Instead of adaptation, one could speak of a *structural coupling* between organism and the environment. This is described by Maturana (1987:101-2) as follows:

> Since a structurally specified system can undergo only state changes specified by its structure, the range of structural plasticity of a composite entity is determined by its structure and not by the medium in which the entity operates and is realized as an entity. The medium can only perturb a structurally plastic system and trigger a change of state that it does not specify. Under these circumstances, the perturbations by the medium act as selectors of the structural transformations of the perturbed entity; and the sequence of perturbations that the medium triggers in the history of the interactions of a given entity acts as a selector of the sequence or course of structural changes that the entity follows in that history. This results in the establishment of a structural correspondence between the given entity and the medium in which it operates, which appears to an observer as adaptation or structural coupling.

Structural coupling explains the "correspondence" between system and environment, or even between different systems, whereby it must be kept in mind that other systems are only environment for a system.[63] Maturana uses the adaptation of the muscles of an organism to the Earth's force of gravity as an example of structural coupling. Gravity is a remarkably constant feature influencing all life on Earth. Gravity selects for bodies with specific muscular capabilities. Those organisms that do not adapt to the force of gravity do not survive. Viability or adaptation on Earth results from the structural coupling between organisms and gravity. For this reason, it would be misleading to use the notion of structural coupling to explain internal differentiation and, thus, internal complexity/contingency in autopoietic systems. The heater and the cooler subsystems within the air conditioning system could be said to be adapted to each other by structural coupling, but this is only possible insofar as they can each be considered autonomous systems in their own right and not merely functional elements of the air conditioner. Are the heater and the cooler really autopoietic, operationally, and informationally closed systems? Do they "adapt" to each other and to the air conditioner? Or are they merely elements of the air conditioning system? The same question can be asked of cells and organs in the body.

Internal differentiation of subsystems cannot easily be regarded as the result of evolution, that is, the self-organization of autonomous autopoietic systems that are then structurally coupled to other components of a system so as to build a systemic unity. Subsystems may, at best, be "semi-autonomous" systems.[64] It would thus appear that subsystems are not operationally and informationally closed but are *functional elements* of some over-arching system. They do not self-organize and

---

63 The correspondence of organism and environment has been discussed since antiquity under the title of "teleology." See for an overview the Wikipedia article: https://en.wikipedia.org/wiki/Teleology. Recently structural coupling has been theorized as a "Markov blanket" under the free energy principle. Friston (2020: 1-2) writes "the free energy principle was introduced to describe how living self-organizing agents have evolved to exist in a confined 'state space' with a bound on their long term entropy.".. which means they evolve "to avoid improbable ... sensory encounters, i.e., those associated with existential risk."

64 Interestingly, Luhmann speaks of social subsystems as "semi-autonomous" and not as autonomous systems. Can systems be "semi" autonomous, semi autopoietic, semi self-referential, semi operationally and informationally closed? We will see later why this is an important question for understanding whether society can be successfully modeled as a system.

therefore do not evolve independently and relate to other body parts via structural coupling. Does the liver, stomach, lungs, or skin self-organize as autopoietic systems and then enter into structural coupling with the heart, brain, etc.? Or are they not much rather *constructed* within the overall system out of the autopoietic operations of the interactions of elements of the system following the genetic and morphogenetic coding of the organism? This is what "autopoiesis" means. This is what it means to say that a system constructs its elements.

Present-day biologists, such as Michael Levin, consider every cell an autonomous system with its own information processing and adaptive operations. Thus, every cell of a multicellular organism would be in structural coupling with every other cell. Of course, Levin (2022) admits that cells form "collective intelligence," which implies operational and informational openness and collective self-reference. According to Levin, bioelectrical networks do this. They could be called a morphogenetic code related to DNA as computer software is related to computer hardware. Out of this collective intelligence arise functional organs such as lungs, heart, skin, etc. To what extent the software of an organism or the morphogenetic bioelectrical networks can themselves be conceived of as autonomous systems structurally coupled to other such software modules within an organism is an open question. How are morphogenetic bioelectrical networks related to each other? Furthermore, the extent to which individual cells that take instructions about what kind of collective entity they should form, heart, skin, bone, etc., can still be modeled as autonomous, operationally, and informationally closed systems is also an open question. After all, they do what they are told to do and not what they want to do themselves. Where is the autonomy? Where are the informational closure and internal information construction? Indeed, cells that beak out of group conformity and go their own way are those cells that Levin calls "cancerous." The model of structural coupling suggests, as Levin does too, that one must think of individual cells in an organism or of subsystems in an organization as autonomous systems. These issues are important because they lay the basis for what it means to speak of a social system which perhaps cannot be modeled in ways similar to biological systems.

There is a great temptation to generalize and apply the ideas of structural coupling and adaptation to every element, for example mitochondria, within an organism, that is, to see them as tiny organisms within organisms, where each is constituted by a system/environment difference, is autopoietic, and operationally and informationally closed. However, we do not find this extension of the theory advisable. On the contrary, cells, tissues, organs, and subsystems should only, in a derivative sense, be said to be "adapted" or "structurally coupled" to the organism as a whole or to each other. Instead, it could be said that they are *integrated* into the organism. There is a difference between adaptation and integration. Can one meaningfully say that the digestive system is "adapted" to the cardiovascular or nervous system within an organism? Could the heart and lungs somehow mutate and put "adaptive pressure" on the stomach? Does an organism's genetic and morphogenetic code represent a kind of ecosystem in which various autonomous systems, over time and under the pressures of variation and selection, have adapted to each? Is it an accident that systems biology drops the word "system" when describing an organism's internal complexity and speaks of "networks of networks"?[65] Why does Levin speak of bioelectrical networks? With regard to later considerations on the relationship between living systems on the one hand and semiotically organized meaning systems on the other, it is important to keep clearly in mind the distinction between structural coupling and integration, or as it is also called *interdependence,* or again, as Luhmann will say *interpenetration*.

A system that allows structural changes to be "controlled" by selection within the framework of a postulated ecosystem is considered to be *viable*. Viability means that the system, however it may change structurally, meets the constraining conditions of the environment. This is called "natural selection." The environment "selects" which organisms are viable and which are not. If the system does not adapt to the environment, it cannot continue its autopoiesis and disappears.[66] When the

65 See Kiani et al (2021).

66 Maturana (1987:107): "In fact, organism and medium are operationally independent systems in their state dynamics, each following its independent structural

structural changes arising from system mutations are so radical that the organization of the system, i.e., its identity, is affected, then we speak of *evolution* or speciation. Evolution is not mere structural change but organizational transformation, i.e., the development or emergence of new forms of systems out of the autopoiesis and operational closure of existing systems. This presupposes that systems can reproduce themselves, which is the very meaning of autopoiesis, and that variation can and does occur in this process. Thus, different species and branches of the tree of life, or phylogenesis, occurs. Governed by the evolutionary principles of *variation and selection,* life explores all possible niches of morphogenetic space. Evolution is the process by which the many different forms of life emerge, transform, and disappear. Evolution is a natural process. If, for example, a genetic engineer creates a new living being in the lab, which is called synthetic life, this is not evolution since the organizational transformation of the genetic code in question did not arise in the process of autopoiesis. Synthetic life is an allopoietic intervention. But after the living being once exists and can reproduce, organizational transformations in the sense of evolution become possible.

1.20 Excursus: Ecology and Sustainability

There is much uncertainty in terminology between the idea of subsystems within systems, the mechanisms constructing internal complexity, and the relation between systems and their environments, including other systems. It would seem that systems theory offers only structural coupling or adaptation to explain how systems relate to anything beyond their constitutive boundaries. Internally, if elements of a system are not defined as autonomous systems in their own right, relations among elements can be modeled as dynamic interrelations. Most systems analysis does this, whether in physics, biology, or society. Almost everywhere the word system is spoken, units are being analyzed into parts that are dynamically interrelated to each other to produce specific

specification. Therefore, appropriate behavior is necessarily possible only as a result of structural coupling, and an organism is therefore either in a state of structural coupling or in a state of dissolution."

operations or emergent behaviors. The modeling is almost always done for purposes of simulation, prediction, and control. This is especially the case where mathematical models are used in engineering to build autonomous machines. Regardless of whether the relations among elements are understood as adaptation or structural coupling, irrespective of whether the system being analyzed is a machine, an organism, an organization, a society, or whatever, the assumption is that the processes of some circumscribed and bounded composite can be described in such a way that simulation and intervention in the operation of the system become possible that allow for prediction and control.

In this context, what do we mean when speaking of an "ecosystem"? In most talk about ecosystems and all of ecology as a science, the environment is falsely conceived of as a system. The various organisms within the ecosystem are then automatically thought of as subsystems within some encompassing "eco"-system. According to Luhmann (1990:21), interestingly, the ecological question does not belong to systems theory at all since it "[makes] the unity of the difference between system and environment the subject, but not the unity of a comprehensive system."

Let us recall that systems thinking became popular at about the same time and is closely connected to ecology. The idea of systems entered public awareness with the first Club of Rome Report, "The Limits of Growth," published in 1972 under the leadership of Dennis and Donella Meadows. The Club of Rome proposed a World Model as the frame used to analyze factors affecting the limits to growth. The model was based on systems theoretical ideas of holism, circular causality, homeostasis, and the interdependent relations among elements leading to emergent properties, properties of the whole that could not be derived from the elements individually. The basic concepts, the methodology, as well as the focus on the interaction of society with the natural environment falsely branded the new science of ecology as a systems science. The systems model became unavoidably related to ecology. The term "ecosystem" expresses this connection, which has continued until today in the idea of "Earth Systems Science."[67]

---

[67] See Steffen et al (2020).

It is interesting to note that the object of ecological investigation, what is described by ecology, that is, the interaction of organisms with an environment, cannot itself be considered a system. It is a fundamental tenet of the systems model that the system/environment difference constitutes a system. The unity of the difference between system and environment cannot itself be a system. For example, a frog may be considered an organic system adapted to life in a pond which is its environment. Suppose the interactions of the frog with the pond are considered to build a system, as talk of an ecosystem suggests. What distinguishes this ecological system from its environment, an environment that it must have to be a system? What is the environment of the pond? If we suppose it is the forest in which the pond exists, the ecological principle of observing all interactions of organisms and their environment together as one unity requires the entire forest to be considered the ecosystem. The whole forest then becomes the system that is being observed. This only pushes the question back. What is the environment of the forest? If the forest is located in the mountains or on a plain, if it is near a big city and thus used as a recreation area, all this becomes the object of analysis. What is the ecological impact of all these factors? Must they not also be taken into account and considered as the larger ecosystem of the pond and the forest? The forest-city or forest-mountain complex becomes the ecosystem. The ecosystem's boundaries must continually be pushed back and enlarged indefinitely until, as is today the case with climate change, the entire planet Earth and beyond the whole universe must be taken account of. Ecology leads directly to Earth Systems Science and beyond to Gaia and the cosmos. Does not the moon, cosmic rays, solar storms, etc., also influence the pond? Is not the ecosystem, in fact, the whole world, as Earth Systems Science in the very name proposes?

Whenever there is a system, the environment, according to systems theory, must be conceptualized as different from the system. How could the system select, relate, and steer elements, refer its operations to itself, and establish and maintain autopoiesis, if there were no difference between the system and environment? Ecology necessarily includes the environment in the system. This means that the system/environment difference is continually dissolved and pushed outward only to be re-established on a higher level. The consequence is that ecology cannot be a systems science because it demands that the interactions between the system and the environment themselves are the object of investigation. Ecology studies what systems science cannot envisage: the unity of the difference between system and environment. For ecology, the environment is unavoidably included within the system, which automatically establishes a new system/environment difference, which is then transgressed until we reach the entire universe, or at least, the planet Earth. As often said, ecology leads directly to the Gaia hypothesis. The world as a whole must, in principle, if not in practice, be included within the ecosystem. This, however, is practically impossible. How can the immense complexity of the world be modeled, simulated, predicted, and controlled? How can models of "sustainability" for the world as a whole be designed? What must be held stable, and what can be allowed to change? How many setpoints must be taken into account? How many different processes? How could the effects of interventions by geoengineering with regard, for example, to the average global temperature, be calculated for all foreseeable outcomes? But this is not only a practical problem that perhaps some supercomputer and unimaginable quantities of data could solve. It is a fundamental problem in theory. According to systems theory, there can be no world system, no system that includes everything. Systemic order is based upon selection, exclusion, and the reduction of complexity. The system is always less than the environment. The ecological model, on the contrary, is based on the inclusion of all factors and the exclusion of nothing. For these reasons, it can be argued that ecology forms a kind of proof case for systems theory. Can the complex interdependencies typical of ecology be modeled as a form of systemic order, and if not, what alternatives are there? Perhaps we cannot limit ecology to the question of life but must include meaning. Perhaps Latour's "critical zone"[68] must include systems of meaning and not only geology and biology. Perhaps Gaia is not a system at all but a network. Perhaps Gaia is not only geology and biology but also culture, society, and meaning.

---

[68] See *Critical Zones. The Science and Politics of Landing on Earth,* Latour/Weibel (2020).

Maybe the ecological crisis is a crisis of meaning, not life. We will return to this question when discussing meaning systems and networks below.

1.21 What is Learning?

The appearance of new behaviors due to structural changes in an autopoietic, operationally, and informationally closed system can be called *learning*. However, learning presupposes that behavior changes are not caused by changes in the organization, i.e., in the system's identity. You may be able to teach an old dog new tricks, but you can't teach an old dog to become a new dog that is a new species.[69] Organizational change is evolution and not learning. Learning is ontogenetic structural transformation and not phylogenetic and evolutionary change. In this sense, one can also speak of "machine learning." This presupposes, however, that the machine can develop new behavior based on its interaction with the environment or data. Simply programming a computer is not machine learning or what could be called "artificial intelligence." Insofar as neural networks can set their weights and create their models to achieve goals, this could be understood as learning. If a human programs the computer, then the behavioral transformation no longer comes from the operational closure of the system but is determined from the outside. In organic systems, one can distinguish between instinctive behavior, where there are no degrees of freedom, and behavior that arises due to the system's interaction with the environment. Michael Levin (2022) has argued that every organism can learn and thus can be attributed some kind of "cognition." Cognition Levin defines as problem-solving, and whenever a system has more than one degree of freedom and can solve a problem in more than one way, it can be said to be able to learn the best way using trial and error. For Levin, this is true of the behavior of single-celled organisms up to human cognition. Levin, and most cognitive science, do not consider the possibility that meaning systems operate entirely differently than biological systems. This confusion leads to many misconceptions about the nature of society

[69] This does not imply that epigenetics and interaction with the environment cannot lead to inherited changes.

and of meaning. It is advisable, therefore, to distinguish different kinds of learning on different levels of emergent order. Machine learning differs from problem-solving on the biological level, which in turn is different from problem-solving on the level of meaning. What they all have in common, from a systems theory perspective, is structural or behavioral change resulting from system operations.

1.22 Meaning Systems, the Semiotic Code, and Cognition

Let us imagine a very complex system. More complex than anything we find in physical or biological nature. Let us assume that the system regards its behavior, i.e., its structural transformations or everything it has learned, as "signs," that is, as something which *means* something. Furthermore, let us assume that the system is self-referential in the sense of being self-conscious.[70] It can be said that the meaning system is a *sign* to itself, or as Pierce once said, "Man is a sign." Furthermore, we will assume that the system can store and retrieve information at will and that it has the ability, perhaps even the necessity, to communicate these stored and retrieved "ideas" to similar systems using language. If all this is assumed, we have again come to a threshold where the previous model of the organism is no longer sufficient to represent the phenomena we are describing.

Just as at a certain level of complexity, the model of the machine had to be replaced by the model of the organism, so the model of the organism

70 Animals of course are also conscious and most probably self-conscious, but they cannot tell us about it so there is fundamental gap between animal consciousness or sentience and the human realm of meaning that can best be characterized as the technology gap. Although life can be equated with cognition and intelligence, these concepts mean something else on the level of meaning. On the level of meaning they mean technology, the "artificial," the mediation of tools in all activities. Again, one must weigh continuity against transformation. Even if all levels of emergent order from the physical to the semiotic are based on one substrate, the substrate is transformed at each level. Despite all continuity, there are undeniable differences in levels of emergent order. The question is whether the differences so outweigh the continuities that the concepts at issue become mere equivocations and the theory loses its generality. We will argue that this is in fact the case.

must now be replaced by a new model that exists on a higher level of emergent order.[71] Just as the genetic code as an organizing principle replaced the mechanical/physical code of the air conditioning system, the genetic code itself must now be replaced by another code that is more complex and variable than life and which can do things with life that life alone cannot do. What kind of code is this? What type of model allows us to replicate the complexity of the above assumptions?

If self-reference occurs as conscious and meaningful self-reference, and learning can be stored, retrieved, and communicated, then we are dealing with what could be called a *meaning system*. A system capable of meaningful operations can be organized by what we shall call a *semiotic code*. It must be emphasized that although semiotics refers to signs, signs are not coterminous with meaning. Long before Homo sapiens, with their big brains and unique linguistic abilities, appeared on the scene, hominins were using tools and creating something that could be called "culture." We are using the term "semiotic," therefore, in a very broad sense that includes, of course, language but goes far beyond that which the discipline of linguistics studies. A semiotically organized system can therefore be called a meaning system or, since meaning and language of some kind are inseparable, a communication system. Thus, communication can become the defining feature of social systems for Luhmann (1986:269):

> A social system comes about when an autopoietic communication context always emerges and demarcates itself from an environment by restricting the appropriate communications. Accordingly, social systems consist not of people, nor of actions, but of communications.

These statements demand an explanation. For Luhmann, the model of a meaning system as a communication system is society. What needs

---

[71] It is estimated that life emerged on Earth about 4 billion years ago. The origin of meaning can be dated to more than 3 millions years ago, that is, with the first stone tools. Homo sapiens, however, with its big brain and linguistic abilities appeared about 300,000 years ago. Meaning is much older than Homo sapiens and also much oder than language.

to be understood is no longer the biologically or genetically organized organism, and not even the human individual with its large brain and the experience of consciousness, which is the subject matter of cognitive science, but instead it is human society, which exists on its own level of emergent order. This evolutionary leap is described by Wilke (1990:27) thus:

> The leap from the realm of necessary stimulus-response linkage to the realm of heightened degrees of freedom and contingency marks a threshold of evolution associated with the concepts of meaning, identity, reflection, self-understanding, and self-thematization.

Accordingly, we must no longer orient systems theory towards biology, and not even towards psychology or cognitive science, but towards sociology and semiotics. This suggestion is highly controversial. That it takes a radical reorientation to describe a system of meaning adequately is not accepted by biologists and most cognitive scientists.[72] These disciplines are forced into a reductionist standpoint by disregarding the leap to a new level of emergent order. Physicalist or materialist reductionism claims only particles, fields, matter, and energy exist, and all other phenomena, including life and meaning, must be "explained" in these terms. Your Grandmother is nothing but atoms in motion. Along these lines, it is supposed that when quantum gravity is finally adequately theorized, this would be a Theory of Everything (ToE).[73] But of course, such a purely physical theory leaves life and meaning out of account and can hardly claim universality. And it cannot explain itself and thus alone on this criterion cannot claim to be universal. But not only the mind is left out of account; physicalist reductionism cannot explain life either. The same game, however, can be played on the biological level. Because of the informational closure of living systems, it is supposed that we, as living systems, are locked within our brains and cannot know the

---

72 The exception that proves the rule is non-Cartesian cognitive science. See for an overview Rowlands (2010).

73 See Wikipedia for an overview. https://en.wikipedia.org/wiki/Theory_of_everything.

world "out there" in the environment. All information is constructed within and by the system according to its own organization. Therefore, the world we know is not the real world but an illusion that is more or less viable.[74] Much of current systems theory is, in one way or another, a form of biological reductionism. Reality is reduced to being an informational construction of the brain or, as is often said, a "model," a "simulation," or even, as Hoffman (2019) claims, an "illusion." This thesis is not new. It goes back at least to Kant and has recently been discussed under the name of "radical constructivism" or "second-order cybernetics."[75] Because of the prevalence, if not omnipresence, of biological metaphors and models in systems theory, there is a widespread tendency to explain meaning systems "biologistically" and thus not to recognize that meaning is a level of emergent order beyond life and, therefore, that meaning systems must be modeled differently from living systems. Since systems theory has been developed almost exclusively as a theory of cybernetic machines and living organisms, it is necessary to address the limitations this imposes on the theory before discussing systems of meaning in their own right.

1.23 Organism or Society? On the Problem of Biological Reductionism in Systems Theory

In contrast to the thesis that meaning must be conceived of as a third level of emergent order beyond matter and life, Maturana, Levin, Friston, Hoffman, and many others claim that the biological model is sufficient to "explain" language and meaning.[76] As Maturana puts it,

---

[74] This is fundamentally a Kantian position to which most cognitive science adheres. See radical constructivism in psychology and epistemology and Hoffman (2019) for a current exposition of this form of biological reductionism.

[75] See the Wikipedia article on radical constructivism and second-order cybernetics: https://en.wikipedia.org/wiki/Radical_constructivism; https://en.wikipedia.org/wiki/Second-order_cybernetics.

[76] The reader should keep in mind that there are many confusing crossovers and interferences between philosophy of mind and its many debates about mind/body dualism, psychological behaviorism, deeply rooted metaphysical commitments to materialism or idealism, and various schools of cognitive science that make the issue discussed here difficult to restrict to systems theory alone.

life can "model" meaning: "Cognition is a biological phenomenon and can only be understood as such" (1985:33).[77] Surprisingly, Maturana (1987:112) openly admits that this approach has nothing to do with meaning or mind in the sense that philosophy of mind has usually understood these terms. He openly admits that the biological model reduces language to "contentless consensual behavior."

> Observers and language have been generated by the linguistic description of living systems and their operations. But in this process, language appeared as a contentless consensual behavior whose validity owed solely to its consistency within the consensus domain.

And even more directly (113):

> The actual operation of language as an ontogenetically established system of coordinated mutual elicitation of behavior between organisms is contentless.

The word "content" refers to semantic meaning. This biologistic view of language deviates so radically from the usual concept of language that Maturana (112) must postulate another, higher domain of operation, which he calls a "consensual domain," to do some justice to the phenomenon of meaning, communication, and intersubjectivity:

> An observer operates in two overlap-free phenomenal domains. As a living system, he operates in the realm of autopoiesis. As an observer proper, he operates in a consensual realm that exists only as a collective realm determined by the interactions of multiple organisms.

The decisive difference between these two non-overlapping domains is that the observer, as a living system, is either individual or collective.

---

77 See also (39): "Living systems are cognitive systems, and life as a process is a process of cognition. This statement applies to all organisms, whether or not they have nervous systems." See also Maturana/Varela (1987:33). Levin (2020) also follows this line of biological reductionism in his theory of life as cognition and Hoffman (2019) takes it to its logical conclusion with his claim that reality is an illusion.

The individual is concerned only with its autopoiesis, but the collective is somehow concerned with the mutual interactions of organisms. What is this collective? As Luhmann will claim when speaking of social systems, it is a special kind of system? Or is it merely a matter of two different perspectives on the same system? Does the collective interaction of organisms create a new and different system on a higher level of emergent order? If one were to consider this "consensual domain" as a system in its own right with its own organization, this would require that one leave the purely biological level and abandon the model of the organism.[78] We dealt with this question above when discussing the idea of ecology. We saw that an ecosystem cannot be a system because it cannot maintain a constitutive difference from an external environment. The same argument applies in the present context. The "consensual realm" of which Maturana speaks as the domain in which the observer supposedly exists is a "collective realm determined by the interactions" of many individual organisms. What informs, relates, and organizes these interactions? What constitutes the unity and characteristics of this collective realm? With regard to ecology, it is uncertain whether Gaia is to be thought of as a super-organism or as something else not describable as a system of the same kind as life. What type of code organizes the consensual domain Maturana ascribes to the "observer proper?"

If we remain within the systems paradigm, we must postulate a fundamentally different kind of system other than autopoietic, self-referential, operationally, and informationally closed systems, or we must abandon the idea that the observer can operate in any kind of special domain. The observer as observer, again, falls out of the theory. This implies that language and meaning cannot be thought of as a system on the biological level of order. Biology cannot account for biological science itself. Within the biological model, meaning becomes an epiphenomenon of life somehow generated by the structural coupling

78 Levin seems to say both at once. The individual cell is at once an autonomous, operationally and informationally closed system and also bound up into a collective that is following instructions from a morphogenetic, bio-electrical network in order to form a multicellular organism, which in turn can be bound up into colonies, groups, and societies. It is unclear whether the units being observed are systems or networks.

of individual organisms or, from the point of view of cognitive science, even individual neurons. It is difficult to explain why individuals should experience this as meaning, as the "world" in which they exist. We find ourselves facing the "hard problem" of consciousness again. One cannot derive meaning from non-meaning, but what we can do, as we argue, is postulate the emergence of a higher level of order upon which language, meaning, communication, and human society appear. Whether they appear as systems is the question systems theory must answer convincingly to maintain its claim to universality. Language is the construction of information by means of communication and not by means of perturbations triggering information construction internal to a living system, even if many organisms somehow do this together.

If organisms are *integrated* into some higher level of systemic order and not merely structurally coupled in some kind of consensual domain, what is this higher domain? How can the "interactions of multiple organisms" give rise to a system of a different kind such that this system is not merely a super-organism? Does life "self-organize" into meaning? Or does meaning "emerge" based on complex forms of life?[79] To answer these questions, Maturana helps himself with metaphors and analogies. First, it is asserted that the observed coordinated behavior between organisms is not something different from human linguistic behavior. For an external(!) observer, what humans do when acting together is no different from what ants or bees do, or at least for such an external observer, it seems to be similar. Bees dance in front of one another just as humans gesticulate and make noises in front of each other. Apart from the fact that neither Maturana himself nor anyone, for that matter, is "external" to language and communication, Maturana goes on to suppose that the structural coupling of behavior among living systems appears to be "like" language (109); thus, a continuity is postulated between genetically and semiotically coded systems:

> When two structurally plastic composite entities interact with each other and thus act as selectors of their respective paths of structural change, mutual structural coupling takes place.

[79] Self-organization is continuous, whereas emergence is discontinuous.

> This results in the state changes of one system recursively triggering the state changes of the other, thus constituting a domain of coordinated behavior between the mutually adapted systems. When this occurs between living systems during their ontogeny, a domain of coordinated behavior is formed that is indistinguishable from a consensus domain that has been established between humans.

And further:

> To an observer who sees two or more organisms operating in a consensual domain, the organisms appear to be operating with consensual representations; therefore, to the observer, a consensual domain functions like a linguistic domain in which the behaviors of interacting organisms create consensual indexes or descriptions of their interactional domain.

Reflexivity is said to bridge the gap between meaningless mutually conditioned behavior on the one hand and meaningful linguistic communication, that is, "consensual representations" or "descriptions" on the other. Meaning appears out of nowhere as "representations" or "descriptions" that the observer somehow knows, or at least supposes, are present for the observed organisms. There are two different things here that cannot be equated. There is the fact that the behavior of a collective of organisms "appears" to an observer as though they were using shared representations to understand each other and coordinate their behavior, and there is the entirely different fact of whether they are actually doing this. It is one thing to appear in a certain way and another to actually be as one might suppose. Where does this knowledge of a linguistic domain come from? What is supposed to be explained by this idea is simply assumed but immediately suppressed and relegated to external structural coupling. According to Maturana, "the interactions within the consensual domain operate as selectors for further structural couplings within the domain." This leads to a "recursion" from operations to operations. From this, meaning and language are said to emerge (109):

> ...then a recursion takes place in which the participating organisms make consensual determinations of consensual determinations. A linguistic domain with such features is a language, and to operate in it is – from the observer's point of view – to operate in a descriptive domain that allows recursive descriptions of descriptions.

That meaning refers to meaning, and communication leads to further communication is not what is observed, but what is experienced. One does not simply observe two people making noises and gesticulating in each other's presence, but one knows they are talking about the weather or who won the football match. Making "consensual determinations of consensual determinations" cannot be understood as structural coupling. Meaning referring to meaning or communication referring to communication is not an example of structural coupling but of the self-referential operations of a meaning system. How the recursion of structurally coupled operations upon themselves magically is supposed to create semantic content, i.e., meaning remains unclear. We speak of magic because one is getting something, meaning, from nothing, non-meaning. How is this to be explained? The model lying behind this sleight of hand is the central nervous system. The brain is supposed to be an operationally closed system of nerve impulses, i.e., "relative neuronal activities" always lead only to other neuronal activities. The system is operationally closed. It operates upon itself "recursively" in that its own neuronal operations lead again to neuronal activity; or, as von Foerster (1993:34) puts it, we are dealing with "the computation of computations." Recursion alone, however, will not do the work that needs to be done here. That nerve impulses operating upon themselves at a certain point of complexity and scale somehow transcend the purely biological level of order and trigger a "higher" level of neuronal networking and processing which is then equivalent to consciousness and meaning, is an interesting and perhaps within the biological model, necessary assumption. It would imply that the lower impulses are somehow meaning-referentially mapped to the higher ones or, as we may also say, semiotically coded. The metaphors of "lower" and "higher" are, of course, misleading. There is no hierarchical structure involved here. There is a leap into another level of emergent order. It is not convincing to suppose that the operational closure of either the neuro-system or the structural coupling of behavior upon itself, that is, mere recursion, can explain the emergence of language and meaning. Where does genetic coding turn into semiotic coding? There seems to be no alternative to postulating the emergence of a higher level of order that is semiotically coded. Maturana (1987:114), and following him, many cognitive scientists are determined to explain the ability to think, i.e., the ability to use meaningful signs linguistically and communicatively, solely based on the operational closure of the nervous system:

> However, as observers, we can always specify a metacognitive domain from whose point of view we are external to it and consider it. We can do this, of course, because in our neural system everything happens in the same phenomenal domain of closed relations of relative neural activities.

As Küppers/Krohn (1992:22) note, semiotic reference or meaning cannot be explained by recursion alone:

> Often...the notion of reference is used synonymously with that of recursion. But while recursion merely specifies that an output of systemic operations becomes its input again, the notion of reference in the narrower sense means that some states represent or refer to others. More sharply, representation or reference is a state that is a representation (a model) of another state.[80]

In conclusion, we can assert that the emergence of language and meaning cannot be explained by means of structural coupling and recurrence of neuronal activity and the resulting appearance to some observer of what is construed as "consensual" behavior. There is no "consensus" here; there is only mutual adaptation. The organisms or nerve cells do not "communicate" or " agree" on anything. Coordination of behavior is not in itself communication or meaningful consensus or agreement.

80 See also Schiewek (1992:154ff), who distinguishes between "operative" and "symbolic" generalization, the crucial point being "that these symbolic generalizations are no longer subject to biological laws...."

Ants, bees, in fact, almost all species in some way coordinate their behavior, indeed, the entire world of life is always a more or less stable state of mutual adaptation. Any "balanced" ecosystem can appear as "coordinated" behavior. But we do not think of this, except metaphorically, as communication and meaning. According to this model, how recursively, i.e., cybernetically interconnected nerve impulses or how a semantically contentless mutually coordinated behavior becomes semantically filled language remains unexplained. That there are cybernetic recursions at mechanical, biological, and semiotic levels is not denied by anyone. But the model of the organism (or of the nervous system) is insufficient to represent the specifically linguistic forms of self-reference and meaningful communication observed and experienced on the level of meaning.

It may well be that, from a particular observer's perspective, structurally coupled behavior looks like linguistic communication. But this does not mean that it actually is language, as is shown, for example, by talk of the "language" of bees or ants, etc. As Jantsch (1987:180) puts it:

> For species representing an early stage of evolution, e.g., insects, the close coupling by physical and chemical processes (such as by chemotaxis) extends further to the association of organisms into societies. But in the later and more complex species, such as mammals and especially humans, the formation of societies is based not only on material processes, such as production and division of labor, but also primarily on mental and spiritual processes, which arguably find their counterpart in neural, chemical, and electrical processes within the organism, but represent a higher level of self-organization.[81]

---

[81] It is equally difficult to understand how "observation" should be possible on the basis of the biologistic model. If observation is seen as a synthesis of "distinction" and "indication" following Spencer Brown (1972), then "indication" is a semiotic term that can only have meaning within a system of meaning. It is impossible to see, as Kneer/Nassehi (1993:97) claim, that a thermostat, which is not part of a meaning system, can "designate" temperature differences. The thermostat designates temperature differences for people who read the device as a sign. For itself, the thermostat does not signify anything and thus it does not observe anything. We will return to the notion of observation below.

We know from Wittgenstein that all behavior is interpretable as language when viewed from the outside, but that is precisely why no external observer can discern language or meaning. Speech, for Wittgenstein, is rule-governed activity, and only those who "know" the rules and are able "to go on" in any communicative situation can recognize language. Merely knowing the rules cannot mean interpreting any particular behavior as rule-governed on the basis of observation, for, as Wittgenstein pointed out, everything can be interpreted as both rule-conforming and rule-violating. Instead, understanding a language means that one knows how to go on, how to participate in a language game. Thus, Wittgenstein can say (1984b: n.102):

> Our paradox was this: a rule could not determine a mode of action since every mode of action was to be brought into conformity with the rule. The answer was: If everyone is to be brought into agreement with the rule, then also to contradiction. Therefore, there would be neither agreement nor contradiction here.

The paradox can be resolved only if the specifically linguistic rules – we can also speak of a specifically semiotic code – function normatively and non-determinatively. This implies that semiotic coding is recognized by being subject to the rules so that one can bump against them and be corrected by other participants in communication. Intersubjective corrigibility is what makes a rule into a rule. Meaning is not a matter of observation but of mutual recognition and participation. Only based on participation in communication can there be "right" and "wrong" in the normative sense. Contrary to this, Maturana/Varela (1987:191) argue:

> Let us be clear, therefore, that the assessment of whether there is knowledge or not is always in a relational context in which the structural changes triggered by perturbations in an organism appear to an observer as an effect on the environment. And the observer estimates the structural changes triggered in an organism in terms of the vortices he expects to see. From this point of view, every interaction of an organism – all its observed behavior – can be evaluated by an observer as a cognitive action.

> Thus, the fact of life itself – the uninterrupted maintenance of the structural coupling as a living being – is nothing else than cognition in the realm of existence. Formulated as an aphorism: Life is cognition.

According to Wittgenstein, only those can determine language and, thus, cognition who can, in principle, participate in communication. But this requires that observer and observed both together already act within a semiotically organized meaning system, i.e., the system reference is a semiotic meaning observable only from the inside and not an organism and not the brain. We find ourselves automatically and necessarily on a higher level of emergent order. In this case, when the observer describes cognition, recognition, meaning, or language, they describe themselves, and thus what they observe is no longer something external to themselves. Observation on the level of meaning systems is always self-observation and self-definition. This is indeed recursive, but it is nothing other than the self-reference of the meaning system. Self-description – we know this from the explanation vs. understanding debate in the methodology of the humanities – is fundamentally different from the objectifying description of external objects and is subject to other semantic and pragmatic rules.[82] The meaning of terms like "observation" and "cognition" changes substantially when it comes to self-observation, that is, to the operations of a meaning system. Although a general systems theory must claim that there are continuities and sufficient similarities so that the same concepts can be used on the biological and the semiotic levels, this, we argue, is not at all certain.

That systems exist only for an observer nevertheless remains valid since observation, as we shall see in a moment, is defined as introducing a distinction. A system can only "exist" when the system/environment difference is applied. The introduction of this distinction is itself an observation. If the operation of distinguishing necessarily implies that the observer distinguishes themselves from the observed, then all observation must be merely external observation. There would be no self-observation. The notion of observation is problematic insofar as it

82 See Apel (1979).

almost inevitably implies the idea of a mere external observation while simultaneously implying self-observation. It is another edition of the subject/object distinction, which entails the possibility that the object can be the subject and vice versa. But if we hold that observation means the articulation of differences, then the classical problems of epistemology concerning subject and object and the paradoxes of the subject that is also object, etc., can be formulated differently.

Meaning systems, which imply cognition, recognition, and communication, cannot be adequately understood by objectifying methods. Indeed, meaning systems do not "exist" for an external observer. Physics and biology do not and cannot explain meaning. The observer, as Heidegger pointed out, is always already "in-the-world" of meaning, indeed, exists as meaning. Furthermore, a theory of meaning systems cannot be based on self-observation as long as observation is understood as objectifying description. With the transition from a physical or a biological theory to a psychological or sociological theory, self-observation becomes self-understanding in the communicative and hermeneutic sense. For this reason, we speak of meaning as a higher level of emergent order. The disregard of this transition in current formulations of systems theory or constructivism explains why these theories are incapable of solving the problem they must solve to justify themselves, namely the problem of meaning. As long as the theory itself is founded on an insufficient analysis of the conditions of meaning – as is the case for all forms of physicalist or biologistic reductionism – it will be self-contradictory.[83]

Furthermore, the fact that there are systems organized on the basis of meaning, i.e., of a semiotic code, is shown by the fact that biological survival, as all those who have died for their beliefs show, becomes subordinate to the principle of meaning. Social phenomena cannot be explained as attempts of organisms to pass on their genetic material or to adapt to an environment to ensure viability. There is no "free energy principle" guiding the emergence of meaning and society. Millions of organisms are sacrificed on the world's battlefields for the sake of

[83] See Bernardo Kastrup for the critique of physicalist and biologistic reductionism, https://www.bernardokastrup.com/.

beliefs, ideologies, struggles for influence and power, etc. These organisms are not concerned with passing on their genes but with passing on symbols such as "freedom," "justice," "the will of God," or "fatherland." Indeed, a life worth living, as Socrates emphasized, is not merely biological but a meaningful life. Meaning makes organisms willing to die for their ideals, religious and ideological beliefs, leaders, etc. The symbolic world of meaning and symbolically mediated group identity is demonstrably "worth" more than mere life. As Wilke (1990:29) argues on behalf of a sociological systems theory, such phenomena cannot be explained biologically:

> At the level of individuals (mental systems), political or religious martyrs of all kinds prove that for people, the preservation of an idea (i.e., a certain form of meaning) can be more important than the preservation of their own lives. The same happens when believers do not touch religiously tabooed animals or plants, although they starve to death. Further examples could be found effortlessly. On the level of social systems, especially 'crusades' of all kinds, point to the fact that groups or whole peoples put their existence at risk in order to preserve or enforce certain contexts of meaning such as religion, political values, or moral postulates. That individual and social levels are as a rule, inextricably intermingled is shown by one of the most ludicrous human inventions, war. For mere ideas (like fatherland, freedom, communism), millions of people let themselves be led to the battlefield. This is no longer biologically comprehensible.

As previously noted, the setpoint of an autopoietic system, its homeostatic parameters, is the preservation of its organization. For organisms, this means the preservation of their life. In the case of meaning systems, or systems organized based on a semiotic code, the system functions to maintain meaning and not to maintain any biological processes. Therefore, we must conclude a higher level of emergent order is involved.[84]

---

[84] In this context, Roth's (1987:269-70) criticism of Maturana can certainly be agreed with: "It is thus factually incorrect to equate life with cognition. What neurons do does not enter into the maintenance of themselves. Simply said: what the muscle cells of my heart and the glandular cells of my liver do has a direct effect

According to Luhmann (1984:141-2):

> Meaning is the very 'substance' of this emergent level of evolution. ... It is completely mistaken to seek a 'carrier' for meaning. Meaning carries itself by making its own reproduction self-referentially possible. ... There are indeed highly complex evolutionary preconditions of meaning formation, but there is no privileged carrier, no ontic substrate of meaning.[85]

The leap from genetically organized biological systems to a semiotically coded system of meaning does not imply, at least this is the theory's claim, that we must leave the conceptual terrain of systems theory and involve concepts outside of the systems model. On the contrary, much more than in any reductionism, the strategy of systems theoretical analysis seems to lie in indicating the level of emergent order and, thus, the framework within which systems concepts can be applied and specified. So, at the sociological level, "every social contact is conceived as a system up to society as the totality of the inclusion of all possible contacts" (Luhmann 1995:33). As is true for the transition from physical to biological systems, so it is true for the transition from biological to

---

on the maintenance of the existence of my organism; but for the activity of many billions of neurons, which are active while listening to Bach's music and at the same time thinking about its complex structure, this does not apply. It is, after all, the characteristic of the cognitive activity of the brain that, if only in some way the continued existence of the organism is secured, it is released from the obligation to promote survival. The autonomy of the brain is quite essentially a release from the maintenance of existence: the brain can occupy itself more and more with things that have only very indirectly or nothing at all to do with survival (or even work against it in the long run). This is precisely the basis of the specific performance of human cognition, namely constitution of reality and thus the possibility of action-planning, i.e., of doing something that is not yet of use to the organism." Roth concludes, "Cognition creates – contrary to Maturana's view – something that does not remain on the same ontological level as autopoiesis." Which of course does not mean that cognition has nothing to do with the organism. The organism together with the nervous system forms the necessary environment of the cognitive system (= sense system). The nervous system, although its operations do not make sense, could be said to be integrated into the meaning system as an environmental condition.

85 Interestingly cybernetics and systems theory have from the very beginning propose the development of artificial intelligence, that is, intelligence not based upon a biological substrate.

meaning systems that all previous basic concepts of system theory are carried over to the higher level. It is also the case, and this must be emphasized, that all concepts developed at lower levels of emergent order must be reinterpreted at the higher level, i.e., interpreted from the code that is the organizing principle of that specific level.

This implies that the meaning system, like all systems – but in its own particular way – can be understood from the functional perspective. Meaning systems, like all systems, function to reduce complexity. As with all systems, meaning does this by constructing a system/environment difference. This constitutive difference allows the system to construct its own elements, which are signs, and build up internal complexity through functional differentiation. Like genetically organized systems, semiotically organized meaning systems are autopoietic in that they generate their own elements and structures from the network of their own elements. They are operationally closed in that they respond to environmental stimuli based solely on their organizing principle, i.e., the semiotic code. Concepts such as adaptation, viability, and learning can also be reiterated at the semiotic level. Nevertheless, the claim of systems theoretical discourse to provide an interdisciplinary terminology through which the difference between the natural sciences and the humanities or between mind and matter can be overcome remains problematic.[86]

Generally, concepts such as "observation," "cognition," "self-reference," "communication," etc., are applied not only to systems of meaning but to all systems. The "problem solving" activity is a case in point when

---

[86] This is evident, for example, in Bateson's (1987:113ff.) systems theory, where there is a tendency to blur essential distinctions. Bateson lists six criteria of mind that are explicitly indifferent across mechanistic, biological, and semiotic levels of emergent order: "1. A mind is an aggregate of interacting parts or components. 2. The interaction between parts of the mind is triggered by differences.... 3. The mental process requires collateral energy. 4. The mental process requires circular (or even more complex) chains of determination. 5. In the mental process, the effects of differences must be conceived as transformations (i.e., coded versions) of preceding events. 6. the description and classification of these transformational processes reveal a hierarchy of logical types intrinsic to phenomena." We will return to Bateson's conception of mind and to this problem of the universality claim of systems theory below.

this is understood to be extended to machines in computer science, cognitive science, artificial intelligence, and even down to single-cell organisms. However, it must also be noted here that all systems theoretical concepts must be understood on the specific level of emergent order that is being modeled by the theory. These concepts have different meanings at different levels of emergent order, and they depend on the model invoked to explain them. "Communication," or "information," for example, mean something different at the physical level than at the biological or semiotic level.[87] We introduce these terms at the level of systems of meaning to indicate that it is at this level that they have their primary sense. "Cognition," "observation," "intelligence," and "communication" are thus primarily terms applicable to systems of meaning, and only in an analogical or derivative sense do they describe biological or mechanical systems. The failure to take into account the different levels of emergent order as the framework for interpreting systems theoretical concepts has been a significant source of confusion and controversy.

---

[87] Michael Levin is particularly liable to equivocate on such concepts as agency and intelligence applying them as he does to physical and biological systems from subatomic particles up to human cognition. These concepts run the risk of becoming so general and abstract as to be useless. What do the "intelligence" of subatomic particles, complex molecules, an amoeba and a human being have in common?

Chapter 2

Meaning as a System: Limits and Possibilities of the Systems Approach in the Social Sciences

The following considerations are conceived within the framework of a general explication of systems-theoretical approaches to the problem of the observer, meaning, and communication. Since Niklas Luhmann can be considered the central figure in this area, we will focus on Luhmann's theory of social systems. Our guiding question will be: To what extent and in what form can we speak of meaning as a system? As stated at the outset, the universality claim of the systems paradigm stands or falls with the extent to which systems theory can explain itself, that is, explain the science of the social and thereby answer the question of what society is. The question of whether meaning can be understood as a system will, therefore, necessarily include a second question: Can the science of society itself be understood as a system?

2.1 Observation

The observer and observation have already been discussed in various ways and different contexts. This was done without a precise definition of the term and to show how a framework for discussing the observer has been inherited from systems models on lower levels of emergent order. The previous discussion of biological reductionism is a case in point. The upshot of that discussion was that the observer could not observe itself as a biological system because observations presuppose meaning, and meaning cannot be derived from organisms' adaptation or structural coupling. Whatever the observer might be, it cannot be an organism. Now is the time to take a closer look at what or who the observer is and what observation means within a broader framework of possible system models.

As stated in the beginning, a system comes into being in the first place because of a difference. Where do differences come from? Who makes the distinctions? This question is answered mainly in connection with the term "observation." The introduction of a distinction is an observation. Systems exist only for an observer. Systems are "constructed" by observation. The observer of physical systems is not a physical system but the physicist, just as the observer of biological systems is not an organism but the biologist. This is so even if the physicist and biologist are personally convinced that they are merely matter in motion or autopoiesis, respectively. We wish to introduce the term "observation" formally and officially at the level of meaning systems. The specific construction of differences that constitutes any system takes place on the level of emergent order of meaning, and this alone should rightly be called "observation."[88] In the most general sense, observation occurs whenever something is distinguished from something else. Without distinction, there is no cognition. Cognition, even in its most minimal or basal sense, implies the distinction between system and environment and some form of self-reference, that is, the distinction between self and other. Only upon this basis can the system "know" what information is for it and how it will be constructed to enable operations. Observation, therefore, is always cognition, and all cognition occurs only through observation. As we saw above, life can only be thought of as cognition in a derivative sense, that is, a sense derived from meaning. Organisms do not think of what they are doing as "cognition," only biologists do. Something can be known "as" something only by being different from something else and from everything else it could be known as. This is a fundamental principle of meaning. Heidegger (*Being and Time* §§32-33) speaks of this principle as the "hermeneutical 'as'" and defines it as the fact that everything that appears can only appear "as" this or that. The hermeneutical as means

---

[88] Thus Luhmann (1990b:683): "In the medium of meaning...being dependent on observational distinguishing and designating is always already built in." Because it is not built in on the levels of physical and biological systems it must be added, often inexplicitly, to the conditions of possibility of systemic order. There arises because of this an ambiguity between ontological and epistemological systems theory. When it was claimed at the outset of our exposition that all systems are constituted by certain principles this description omitted the observation. Now we can add to these general principles the constitutive function of observation since all systems are constituted on the level of meaning.

that Being is contingent; everything can be otherwise. If something can only appear "as" this, it could also appear "as" that. Being is contingent since it refers to what it is not and what could be. We know *that* something is and *what* something is only if we know what it is not or what it could be, i.e., if we know what it is different from. Luhmann (1990:44-5) speaks of a "difference technique" in this context:

> Even within meaning-processing systems, as in living systems, one's own autopoiesis must be secured as a matter of priority. That is: the system exists only if and as long as the meaning-processing of information continues. The structural technique that makes this possible can be called difference technique. The system introduces its own distinctions and, with the help of these distinctions, records states and events which then appear as information for the system itself. Information is thus a purely system-internal quality. There is no transfer of information from the environment into the system.

Citing Maturana, Luhmann (1995:36) defines observation at the level of general systems theory as an empirical/factual operation of a system and defines it as the "handling of distinctions." In this, Luhmann, like Maturana, has been influenced by George Spencer Brown's Laws of Form (New York 1972).[89] Spencer Brown places an arbitrary operation of distinguishing – "draw a distinction" – at the beginning of everything. The distinction, however, must then be coupled with an "indication." Who or what is drawing the distinction and making the indication is not designated. There is no system reference carrying out the operation of distinguishing. The operation can be attributed to an "observer." It can be understood as an operation of "observing" only because of other distinctions, e.g., the distinction between operation and system and between observing, observer, and observed. Of course, one can speak of observing only if an observer and observed are thought together and if observing as an operation is distinguished from what is operated upon, and operations are distinguished from states, and so on. Again, it must be kept in mind that these concepts are meant to be generally applicable

89 For the discussion surrounding Spencer Brown see Baecker (1993a, 1993b).

to all systems and not only meaning systems. Nonetheless, only in a derivative sense can observation be attributed to biological systems. The differences that biological systems introduce and the acts of observation that construct these differences are only available on the level of meaning. It is meaning that decides to whom or what these terms should be applied. After the emergence of meaning, all biological and physical phenomena are integrated into meaning. There is no particle or field, no atom or molecule, no organism that is not contextualized and constrained by meaning. The semiotic code is more complex and variable than the physical or biological code and integrates these lower levels of order into it. As Heidegger would say, Being is meaning.

It can now be asked what comes first, the distinctions or the operations that are said to introduce the distinctions in the first place. Without operations of observation, there are no distinctions, and without distinctions, there are no operations. What comes first? Is there first an observer or an observation? To answer this question, one could argue that selection, relationing, and steering constitute any system as a system and construct the operations and elements. Observation, therefore, is an effect or result of system formation. First, the system is there, and then it can perform operations of observation. On the other hand, one could ask how the foundational system/environment difference could arise without observation. Nothing outside the system could be thought of as a source or cause of observation before the system's self-organization through selection, relationing, and steering. According to systems theory, complexity can only be reduced by these principles. The moment self-reference is constructed using the distinction between system and environment and the closure of system operations, the notion of the observer who operatively introduces distinctions becomes available. Identifying an observer is a product of distinctions, such as that between observer and observed, operation and state, system and environment. Therefore, the question arises of under what conditions (i.e., distinctions!) are distinctions attributed to the operations of an observer who itself is constructed on the basis of distinctions?

The fact that systems theory, in the form proposed by Heinz von

Foerster, Maturana, Varela, and Luhmann, conceives of distinctions as operations and then defines operations as the "factually" accomplished observations of an observer does not necessarily follow from the assumption that the constitution of meaning occurs through constructive articulation of differences. Instead, it represents a momentous decision to base specific differences on all others as "guiding differences." It is in the tendency of this decision to conceive of meaning as either the product of an observer who has become a "subject," standing at once and consequently paradoxically outside and inside the boundary of meaning, or as the product of an undifferentiated and therefore unthinkable, anonymous operating. If systems theory, on the one hand, does not want to hypostatize the meaning system into a "subject" in the sense of modern epistemology and metaphysics, and/or, on the other hand, fall prey to the irrationalism of a blind primal act in the sense of Spencer-Brown's "draw a distinction," then it must find a plausible model for the articulation of differences, or in other words, it must offer a good explanation of the origin of meaning.

2.2 Self-Reference

If the system applies the difference technique, i.e., the distinguishing operation of observation, to itself, then we have self-observation or *self-reference*. An autopoietic and, therefore, recursively operating meaning system must do this; otherwise, it would not be able to know which events are its own preceding operations and thus connect following operations to them in a continuous chain of meaning. But what "self" does self-reference refer to when everything, including the environment, has meaning? There is no system that is meaningful on one side, a kind of subject, and the world that is not meaningful on the outside as Kant was forced to postulate to locate meaning within a knowing subject. Those who criticized Kant's dualism argued that the subject must distinguish between itself and the environment (the thing in itself) within itself. For meaning systems, the environment is constructed within the system as the necessary reference to an "other," enabling reference to "self." Self-reference and other-reference are mutually dependent. There is no self-reference without reference to an other, which is not the self. The system must be able to observe itself, i.e., to distinguish itself. But it must, therefore, differentiate itself from what it is not. Like all systems, it distinguishes itself from its environment and constructs itself by replicating or modeling the system/environment difference within the system. Thus, Luhmann can say (1995:36-7), "self-observation is the introduction of the system/environment distinction within the system, which constitutes itself with the help of that distinction." In this way, a self-perception or an "identity" is constructed.[90]

Luhmann (1995:33) defines the term "self-reference" as a form of self-constructing, an operation that constructs unity by means of distinction.

> The concept of self-reference designates the unity that an element, a process, a system is for itself. 'For itself' means independent of observation by others. The concept not only defines, but also contains a significant statement, for it maintains that unity can only come about through a relationing operation, that it must be produced and that it does not exist in advance as an individual, as a substance, or an idea of its own operation.

There are three different forms of self-reference, depending on whether it is elements, processes, or the system as a whole that is referred to as a self in each case. At the level of system elements, there is a minimal form of self-reference, which Luhmann (443) calls "basal self-reference:" "Basal self-reference is the minimum form of self-reference without which autopoietic reproduction of temporalized systems would be impossible." To produce itself autopoietically, the system must recursively connect its operations to its own operations. Metabolic processes in an organism, for example, must connect to other metabolic processes such that the same processes are autopoietically re-produced and homeostasis can be maintained. In this sense, all autopoietic, operation-

---

[90] See Luhmann (1990b:482): "This designation (observation, description) of the system by the system itself we wish to call reflection, and to note the difference from a mere generation of the unity of the system (seen by an external observer), in the case of reflection we speak not of unity but of identity."

ally closed systems can also be called self-referential:

> One can call a system self-referential if it itself constitutes the elements that compose it as functional units and runs reference to this self-constitution through all the relations among these elements continuously reproducing its self-constitution in this way. In this sense, self-referential systems necessarily operate by self-contact; they possess no other form of environmental contact than this self-contact. (1995:33)

The difference between systems of meaning and biological or mechanical systems is the way the system "refers" to itself. It is not merely that operations are referred to system operations but that there are different system "identities" to which operations are referred. The system builds up internal complexity by distinguishing levels of self-reference. Basal self-reference directs meaning to meaning. On a different level, agency or ascription of meaning to a subject of some kind is another form of self-reference. And finally, the system as a whole can be referred to. This is the level at which the system-environment difference is thematized. As soon as we are concerned with system processes as interactions with an environment, that is, with the re-production of the system/environment difference within the system, we are dealing with the specific form of self-reference peculiar to meaning systems as totalities. The environment of a meaning system, this was the critique of Kant's dualism, cannot be outside the system but must be a form of meaning within the system.[91] Meaning systems construct themselves by reintroducing the system/environment difference into the system. Following Spencer-Brown, Luhmann (1990b:83) speaks of "re-entry." Thus, the

[91] Hoffman (2019) for example buys into Kantian dualism and biological reductionism by insisting that internal information construction implies that "reality" is an illusion. Because illusion, or "simulation," is a comparative concept only meaningful in distinction to reality or to something that is being simulated or modeled, there must be a reality in itself, a noumenon, which we cannot know, but can still somehow simulate and model. But if we can't know the original, what sense does it make to talk about a simulation or a model? How do we know if it is a "good" simulation or a bad one? Hoffman can refer to natural selection and claim that as long the system exists it must be viable and that's all we need to know. This is, however, a tautology, that asserts only that what exists exists.

system refers to itself as a whole, which it can only do by simultaneously "referring" to the environment, which must also have meaning and is, therefore, inside the system and not outside. As a reference within the system, the environment has meaning. Paradoxically, the environment is included in the system and thought of as "belonging" to the system. This unusual inclusion of the environment within the system causes problems for the articulation of the system's *primary code*. From a logical perspective, pure self-reference is tautological: A is A. To articulate further distinctions and not cause the meaning system to be eternally locked into repeating A is A, the original tautology must be de-tautologized: A is non-A. De-tautologization, however, creates a paradox. This paradox must be de-paradoxed to allow the construction of further distinctions and thus allow further operations of the meaning system. This process requires a special "semantics of de-paradoxification" typical of articulating the primary code of a meaning system. We will have occasion to discuss these issues and what a "primary code" means in detail below. The point to be emphasized now is that self-reference and other-reference (environmental reference) are simultaneously and interdependently constructed within the system:

> ...unlike nervous systems, structures and processes that employ meaning can include system boundaries and environments, which take on meaning within the processes of a self-referential system (not: in themselves!), so that such systems can operate internally with the difference between system and environment. For all internal operations, meaning enables an ongoing reference to the system itself and to a more or less elaborated environment.... (1995:37).[92]

[92] See Luhmann (1984:601): "We will speak of reflection when the distinction between system and environment is at the basis. ... In this case, the self is the system to which the self-referential operation ascribes itself. It takes place as an operation by which the system designates itself in distinction from its environment. This occurs, for example, in all forms of self-representation...." For a formulation that emphasizes external reference, see Luhmann (1990b:306): "But since systems that are operatively bound to meaning and closed on this basis cannot find the notion of a meaning-free environment – they would have to be able to do this themselves – they can only refer to their environment internally, that is, only in the medium of meaning."

The reflection of the system is necessarily totalizing and universal because observation cannot go beyond or behind the difference between system and environment or subsume it under differences that somehow lie further back. Thus, the meaning system constructs itself equiprimordially with constructing a "world view." Heidegger (*Being and Time*) speaks in a similar sense of a non-subjective "understanding of being," which simultaneously discloses itself and the world as a whole. The understanding of being forms the horizon of meaning of every self-reference, and one cannot go beyond it. As that being which is concerned with itself, Dasein must be able to refer to itself and to what it is not at the same time. There is always self-reference and other-reference within the one all-embracing horizon of Being. However, to be able to refer to oneself, one must already have interpreted oneself as something, i.e., interpreted oneself as a being of a certain kind. Everything else is interpreted as being of another kind. In this respect, the phenomenological descriptions of "world" that the early Heidegger employed are not irrelevant to systems theory. Luhmann, interestingly, does not refer to Heidegger to explain the concept of meaning but, as we shall see in a moment, to Husserl. We will ask whether a certain distortion or limitation of the concept of meaning in systems theoretical discourse might not be due to this dependence on Husserl instead of Heidegger.

2.3 Meaning

A system of meaning constitutes itself, that is, internally differentiates itself, including the system-environment difference, through successive operations of distinction, i.e., observation and self-observation, in such a way that a "world" is opened up, a world in which the system itself occurs as something distinguished through self-reference/reflection. This entire web of differences, distinctions, referents, and relations, which from a systems theoretical point of view has the function of reducing complexity in a certain way, is what we call meaning.

In previous discussions, we have spoken several times of meaning as the organizational form or code of a specific level of emergent order.

We have generally noted that meaning as a fundamental concept in systems theory is not to be understood, as is usual in the philosophical tradition, as a property or product of mind or of a subject – regardless of how mind or subject might be conceived. Within systems theory, mind is conceived of functionally as a reduction of complexity through a system/environment difference constructed by internal differentiation of the system. Subjects are one form of self-reference of the meaning system. What is peculiar about meaning is that the meaningful distinction between self and other is drawn within the system as a sign. The code, the organizing principle of meaning, thus maps the system semiotically.

There is no meaning without signs or symbols in the broadest sense of these terms. Or if there should be such, we could neither talk about it, think about it, nor know it in any way. As the idealist critique of Kant argued, we can talk about what we cannot know and, therefore, it is known. The sign, therefore, has no other, no unsignified, no other from which it could be distinguished and in terms of which it could be defined. The function of the sign consists precisely in determining the "what" of possible thinking and speaking. Meaning differs from what is often referred to as "sentience" or what is typical of animal "consciousness" or "cognition" as described, for example, by Maturana, Levin, Hoffman, and many others. If biological systems theory equates cognition with life, then signs are regarded as mere signals or triggers without semantic reference.[93] A single-celled organism may be processing information from its environment to sustain its autopoiesis, but it is not constructing meaning. If we do not want to blur the difference between semiotic and genetic coding, and thus the difference between signs and signals, we must hold that at the level of meaning systems, observing or distinguishing is necessarily accompanied by "signifying." Distinctions that are not or could not be named, or as Spencer-Brown would say, "indicated," are not meaningful distinctions. Naming is meaning-making. In terms of systems theory, we can say that "language" reduces or organizes the infinite complexity of the primal substrate into

[93] It is at this point that much current discussion on "free will" is generated.

meaningful contingency. The "hermeneutical 'as'," which Heidegger understood to be how Being appears as meaning, implies that whatever appears can appear otherwise; that is, it appears against an infinite horizon of possible interpretations but cannot appear as anything other than meaning.

The "subject" or bearer of the operation of introducing differences and constructing relations is the system itself, which organizes itself "autocatalytically" as a system of meaning. It articulates a primary code and opens up a symbolic order. The opening up of a symbolic order – called "world" for short – is done by communication. We thus equate semiotic complexity reduction with communication. Meaning and communication are closely connected. Communication "makes" meaning. Without communication, there is no meaning. Accordingly, the conditions of meaning will have to be sought in the conditions of communication, not the operations of the brain or some mysterious entity called "mind." Systems theory is not a theory of mind in any sense in which this word has been interpreted throughout Western philosophy and especially in European modernity since Descartes and, above all, in contemporary cognitive science. Systems theory does not talk about subjects and objects and wonder how they epistemologically relate to each other and what they might ontologically be as entities.

These definitions and assumptions mark our position regarding the status of systems theory in the social sciences. They allow us to conceive of systems of meaning as systems of communication and not merely as cognitive systems or psychological systems, which inevitably brings with it traditional assumptions about human individuals that do the thinking, know the world, speak, act (freely or not), are morally responsible (or not), are deluded (or not), are living in a simulation (or not), etc. Systems theory, we propose, is not bound to the assumptions of humanism, Western philosophy of substance, or cognitive science. Its potential goes far beyond the predominantly modern Western framework in which it is mainly used. This opens up an approach to analyzing systems of meaning free from the presuppositions of Western modernity. However, this thesis is not uncontroversial. Luhmann himself seems to uphold traditional views by attempting to distinguish between meaning and communication, with communication representing only one particular form of meaning. For Luhmann, there are two different and mutually exclusive kinds of systems on the level of emergent order of meaning: psychic systems or human individuals and their brains, and social systems constituted by communication and the exclusion of human individuals and brains. In what follows, we will attempt to argue for a systems theoretical analysis of meaning that places itself squarely on the side of language and communication and leaves non– or pre-linguistic psychological systems to the neuro and cognitive sciences, that is, to lower levels of emergent order than meaning.

2.4 A Critique of Luhmann's Theory of Meaning

Luhmann regards meaning as the basic concept of a systems theoretical approach to sociology. Despite the sociological perspective of his theory, he defines the concept of meaning so broadly that meaning also appears to be constitutive for non-social and non-communicative systems, that is, for "psychic" systems. Luhmann is to be credited with clearly assigning meaning to a level of emergent order in its own right beyond life. His theory of meaning as a level of emergent order beyond the physical and biological levels thus claims general epistemological and philosophical relevance. Meaning is not something that living systems do. Brains don't make meaning. Meaning alone makes meaning, and it makes nothing else than meaning, just as life operates to maintain life and nothing else. Everything that has long been considered under the term "mind," that is, rationality, thought, consciousness, experience, emotions, beliefs, etc., in short, what traditionally characterizes the human as human is now to be conceptualized as a form of systemic order. Nevertheless, systems theory inherits the legacy of philosophy, its questions, assumptions, and even its methods to a certain extent. Therefore, there is an ambiguity in the theory, which results from the fact that traditional ways of conceptualizing meaning derive from and are unavoidably associated with the subjectivist ontology and epistemology of modern philosophy of mind.

From Descartes to Husserl, to talk about meaning meant to talk about a conscious subject. Even if this subject, as in the case of Hegel, is no longer the human being, but *Geist*, the world spirit, and as in Kant, no longer empirical but transcendental, that is, no longer merely individual but universal. It was generally accepted and, for the most part, still goes without saying that whatever is referred to by language is an object of subjective intentionality, an act of knowing by a cognitive agent. Within this frame of thought, the idea of a "system" of meaning almost inescapably implies a subject, literally as that which "underlies" all thought and meaning. In modern thought, this subject was "transcendental" if it preceded experience as a condition of its possibility. Or it was "empirical" if it was the object of an experience of self, a self-reference, a person with their own inalienable experiences. Starting from Descartes' mind/body dualism, modern philosophy understood mind to be either an individual human being or, as in cognitive science, a brain. In its transcendental form, mind was a universal principle of reason, or a collective intelligence, a super-self like God, which grounded all beings. If it should happen that individuals or brains communicate with each other by means of language, then language was conceived of as a cognitive tool. Subjects could use language to express their mental states or choose not to communicate and remain within a purely subjective world of inner experience.

Communication, therefore, was almost inescapably understood to be the act of a subject and attributed to a subject who is not only a subject of free will (someone who makes choices, for example, the choice to speak or not) but who is also in some way the source of meaning, the instance that, as Husserl puts it, "constitutes" meaning in intentional acts. The consequence of this tradition is that meaning is inseparably associated with consciousness, which in turn is inseparably associated with an "I," an "identity," a subject of meaning. The materialists claimed that the I is an epiphenomenon of neural activity and, therefore, a kind of illusion. In contrast, the idealists struggled to maintain and distinguish the universal consciousness of reason from the particular consciousness and identity of unique and often irrational human individuals. When systems theory proposes to conceptualize meaning in its own right, it must carefully maneuver through the obstacle course of the modern philosophy of the subject. This is exactly what Luhmann attempts to do. Below, we will examine whether he succeeds or fails and what the consequences of this attempt mean for the systems paradigm.

The dependence of systems theory on modern philosophy of mind can perhaps most convincingly be illustrated by the fact that Luhmann (1995:60ff) bases his theory of meaning on the phenomenological analysis of meaning found in Husserl's theory of consciousness. This momentous decision arises not from any internal necessity of the systems model but from the above-cited fact that modern philosophy of mind is inseparably associated with consciousness and the idea of the subject. For modern philosophy, there is no meaning that is not the meaning of a subject. All consciousness is consciousness of something by someone. However, neither consciousness nor the subject are fundamental concepts of systemic order. Despite this, Luhmann does not attempt a theory of meaning based on systems theory. Instead, he tries to exploit an ambiguity concerning the status of the subject in Husserl's thought in order to propose two different kinds of meaning system. Husserl's phenomenology distinguishes between the empirical and the transcendental ego, or what could be understood as the individual mind, and the collective or universal mind. Luhmann transforms the transcendental ego into society and communication, while leaving the empirical ego to psychology. He then bans empirical individuals into the environment of the social system, leaving society to consist exclusively of communication and not consciousness. Let us take a closer look at this theoretical maneuver.

Luhmann (1995:60) begins by pointing out that Husserl describes meaning as a totality or "world" of interconnected references.

> The phenomenon of meaning appears in the form of a surplus of references to further possibilities of experiencing and acting. Something is in the focal point, in the center of intention, and all else is indicated marginally as a horizon for an 'and-so-forth' of experience and action. In this form, everything that is intended

> holds open to itself the world as a whole, thus guaranteeing the actuality of the world in the form of accessibility. Reference actualizes itself as the standpoint of reality. It refers, however, not only to what is real (or presumably real), but also to what is possible (conditionally real) and what is negative (unreal, impossible).

Overlooking for the moment the reference in this citation to "experience and action" and "intention" to which we will return, if reference can be said to "actualize itself," then it is not necessarily something that is constituted by a subject, that is, by something other than the system of references as such. In other words, the system is the actor. If the system is the actor, this implies that consciousness and meaning are not necessarily the same thing, at least, as long as we are talking about a personal consciousness and not some kind of world-consciousness. Consciousness seems necessarily to imply a subject that is conscious of itself. This is the subject of intentionality, the subject that knows that it is aware of an object. The object is the reference, which appears against the horizon of further possible references. Meaning is not in the mere fact of intentionality. Intentionality means that all consciousness is consciousness of some object that is necessarily "associated" with other possible objects in a world or against the horizon of a world. It is these associations or references that constitute meaning. Meaning is not some particular thing of which we may be aware of, but necessarily the horizon of other possible things which could become "thematic" for us, that is, objects of other intentional acts.

Luhmann cites Husserl's *Ideas of Pure Phenomenology and Phenomenological Philosophy I* and *Experience and Judgment*. The upshot is that any analysis of meaning always presupposes meaning and thus must articulate itself as self-description. For this reason, Luhmann thinks he must start from phenomenology, an unprejudiced description of experience. But is the experience of meaning the experience of a self-conscious subject? Or does it merely show an open horizon of references, a world in which subjects may also appear?

> Meaning refers to further meaning. The circular closure of these references appears in its unity as the ultimate horizon of all meaning: as *the world*. ... Husserl outlined this situation in the metaphor of the 'horizon,', without thoroughly analyzing the self-reference of all meaning. All proofs of this statement must already presuppose it; they have no other way of operating than by reflecting on the world within the world. We begin by phenomenologically describing the experience of mean and the nexus of meaning and world constituted simultaneously, not basing this description on the underlying existence of an extramundane subject (which everyone would know existing in oneself as consciousness) but understanding it as the self-description of the world within the world. (Luhmann 1995:69)

The "subject" of this description is what Husserl called the "transcendental Ego." It is a pure consciousness that is somehow, as already for Kant, related to the empirical ego, the self that I know myself as an individual to be, but is nonetheless not a person, an individual and thus coterminous with the entire world. There is no necessary implication that we can have meaning without conscious subjects, only that when meaning emerges, it also has consciousness within it. Meaning comes first and is foundational, whereas subjectivity arises within meaning, as we shall see, as a form of self-reference. So Luhmann can write (1971:28):

> The concept of meaning is to be defined primarily, that is, without reference to the subject, because the latter, as meaningfully constituted identity, already presupposes the concept of meaning.

In this definition of meaning as a system of mutual references against a horizon of further possible references, if there is a subject, then it is one reference among all the others. In this way, Luhmann attempts to make the phenomenological concept of meaning fruitful for non-subjectivist oriented systems theory. His program is to reinterpret the concept of meaning and, thereby, the entire modern subjectivist philosophical tradition in terms of five fundamental concepts of systems theory:

1) reduction of complexity

2) system/environment difference

3) self-reference

4) autopoiesis

5) operational closure

Reduction of Complexity: Functionally, as with all systems, meaning is concerned with reducing complexity. The event or operation of meaning *selects* something as an intentional theme or object from a horizon of possibilities. When we talk about something, we mean this or that within a specific context or frame of relevance. We thereby select something particular out of all possibilities. The phenomenological concept of intentionality, which is the act of a conscious subject, is here reinterpreted in terms of the systems theoretical concept of selection, which is an organizing process of a system. When something is selected as a content or object of intention by observation and distinction, all that the object is not is also simultaneously co-thematized. If a hammer, to use an example from Heidegger, is selected/thematized, then at the same time, all that is not a hammer becomes a co-thematized horizon, i.e., there arises a *relationing* of references or possibilities of selection referring to all the other tools, to the workshop, to the various materials that are to be worked upon, to construction plans, to the people or machines doing the work, even to the possibility of not having to work at all, and so on indefinitely. These relations are directed or *steered* toward possible goals. Every event of meaning involves, therefore, selecting, relationing, and steering. These system-building processes reduce complexity in a certain way. World, that is, the meaningful world, as a system, is always a form of reduced complexity. We encounter and deal with things like hammers in a world where hammers, tools, workshops, tables, etc., all have their place. Without the totality of these references, a hammer could not appear "as" a hammer at all, and no meaning would be possible.

The distinction between actual and possible and the accompanying associations and relations can be understood as a *reduction of complexity* in two ways. First, specific contents are selected and explicitly thematized, and second, specific other contents are accordingly related as possibilities for further references. Beyond this, many other possibilities are excluded as irrelevant.[94] Not everything possible in the world is included in the horizon of a typical workshop where things like hammers have their natural home. Unless I am a physicist, when using the hammer, I do not consider the electromagnetic field and the strong and weak forces within the nuclei of atoms making up the hammer. Not everything in the world appears relevant for further distinctions and observations. Only those objects and references compatible with a world in which hammers exist in certain ways and for certain purposes appear. Thus, the thematization or observation of an object, such as the hammer, takes place within a certain horizon of possible distinctions and observations. This horizon and all its possibilities is not something an individual subject creates. It is what the system does, and this horizon appears as historically given. A carpenter's workshop in a medieval village has a different horizon with different references and possibilities of action than a modern furniture factory. Indeed, the language that at any time is spoken and the entire social order in which this language is spoken is not created by the individuals who speak it and act within the possibilities that the particular world they live in offers. Heidegger called this fact of human existence "being-in-the-world," whereby the "world" is always a historically given world into which Dasein is "thrown." Luhmann interprets this as the result of systemic operations of selection, relationing, and steering, which reduce complexity. The at any time relevant horizon of meaning arises from the transformation or reduction of indeterminate complexity, wherein anything and everything is possible, into the determinate complexity of a historical world. Out of the chaos of all possible observations and distinctions, only some are selected:

With each and every meaning, incomprehensibly great

[94] This is known as the "framing problem" in AI, cognitive science, and philosophy of mind. See for an overview https://plato.stanford.edu/entries/frame-problem/.

> complexity (world complexity) is appresented and kept available for the operations of psychic and social systems. … …every specific meaning qualifies itself by suggesting specific possibilities of connection and making others improbable, difficult, remote or (temporarily) excluded. (Luhmann 1995: 60/61)[95]

System/Environment Difference: If meaning reduces complexity, it can be seen as a particular way of drawing the system/environment difference. The system-constitutive difference between indeterminate complexity/contingency and determinate complexity/contingency is accomplished in systems of meaning by means of those special distinctions that constitute observations of others and observations of self. As we have seen, all autopoietic, operationally closed systems must refer their operations to themselves to ensure that what they do is directed toward ensuring their autopoiesis. Autopoiesis means that a system must operate to ensure its own operations. Unlike genetically coded systems, where the constitutive difference between system and environment is an actual physical boundary such as a membrane, skin, etc., and the internal relationing of the elements determine system operations which aim at maintaining and reproducing these structures via autopoiesis and homeostasis, semiotically coded meaning systems do not have "real" boundaries. Just as the boundary of a living system must be something living, the boundary of a meaning system must be meaningful. Contrary to living systems, however, where the boundary is physical and physically separates the system from an external environment, the boundary of a meaning system is inside the system since it is meaningful, and therefore, distinguishes the system from an equally meaningful environment, that is, an "internal" environment. For meaning systems, the system/environment difference is itself something meaningful, and thus within the system and not a real boundary separating the system from an external environment. Outside the meaning boundary, there is nothing, not even nonsense, because even nonsense has some meaning; otherwise, we couldn't talk about it. Nothing, and

[95] Again within cognitive science and AI this is discussed in terms of what is called the "framing problem." See Anderson et. al. (2022), who argue that cognition can be understood as a solution to the framing problem.

not even "nothing," is excluded from a meaning system. For meaning systems, it can be said that the outside is necessarily inside. The world horizon of possible references is infinite. Who can say what is not in some way possible or thinkable? The very idea of the horizon suggests endless extension and not bounded totality. We cannot see beyond the horizon. Nevertheless, the world of meaning is not infinite complexity. Every system must construct its relevant environment. Our air conditioner existed in a world of only a few relevant temperature differences. For the frog, only quickly moving black patches across the visual field are relevant information, not the concert in the nearby park. Meaning systems always emerge together with an environment that is a structured complexity, if for no other reason than that the environment must also have meaning and thus be "contained" within the system. Every religion, worldview, or philosophy claims to circumscribe all that exists conceptually and to understand the world in its totality.

> Meaning becomes [for meaning systems AB/DK] the form of the world and consequently overlaps the difference between system and environment. Even the environment is given to them in the form of meaning, and their boundaries with the environment are boundaries constituted in meaning, thus referring within as well as without. ... The system's differentiation with the help of particular boundaries constituted in meaning articulates a world-encompassing referential nexus, with the result that it becomes possible to ascertain when the system intends itself and when its environment. But the boundary itself is conditioned by the system, so that the difference between the system and its environment can be reflected as a performance by the system, that is, thematized in self-referential processes. (1995: 61/62)

It is thus a unique characteristic of meaning boundaries that they include in themselves all that is outside. "Nonsense" also makes sense, for otherwise, we could not think or speak about it.[96] Meaning,

[96] It is interesting to note that all religions and cultures have a problem with "other" Gods, beliefs, ways of life, etc. They are usually judged negatively as "unbe-

according to Luhmann (1995:60), is a concept "without difference," that is, a concept "which includes itself" and therefore cannot be defined by opposition since it contains all differences and negations within itself. This generates paradoxes. How can something be A and non-A, meaning and non-meaning? The paradox, however, and the resulting need for "de-paradoxing" are typical of systems of meaning.

Self-Reference: These considerations lead to the third way in which systems theoretical foundational concepts apply to systems of meaning: self-reference. Not only does the emergence of meaning imply reducing complexity and system constitution, but meaning is also self-referential in various ways. First, every actual unit of meaning, i.e., every thematized content, in turn refers to itself as a possibility, i.e., by thematizing something, all other possibilities, as well as it itself, remain in view. The hammer, for example, can be thematized only if it can become thematized again with all the associated references to nails, wood, etc. If one turns away from the hammer to get some nails, the hammer doesn't just disappear. The meaning unit is always theme and horizon together. In short, we can talk about the hammer only if we can talk about it again and in many different ways. Translated in terms of systems theory, this means that every distinction occurs as a possibility within the horizon in which it itself is contained:

> Every intention of meaning is self-referential insofar as it also provides for its own reactualization by including itself in its own referential structure as one among many possibilities of further experience and action. (Luhmann 1995:61)

The possibility of mere reactualization is one form of self-reference, which Luhmann calls basil self-reference. Meaning, however, is also self-referential in the sense of being a thematic content itself. Self-reference of this kind can be called "reflection" since it reflects upon itself as in a mirror. The system "sees" itself and thus constructs an identity. The hammer, of course, is not conscious of itself. But the meaning system

lievers," "ignorant of the true God," "abandoned by God," "primitive," or their followers are considered not to be human at all. See Krieger (1991b/2006) for a discussion.

that selects the hammer becomes, in a certain sense, the subject of this action, and thus something that the system also thematizes. After all, when hammers are being used, someone, if even a robot, is using them. When the system selects the hammer, it simultaneously selects itself, but not as in basil self-reference, when it refers to the hammer, but in the ascription of agency. This fact has generated much theoretical effort in modern philosophy. For some, such as Descartes and Kant, this reference was interpreted as a reference to the ego or the *res cogitans* in Descartes' terminology. But what was this thinking subject? Was it the personal ego or something greater, something universal just as reason is universal and not a property of me? The "I" that, according to Kant, must necessarily accompany all perception and knowing was the transcendental ego and, in some sense, the empirical ego as well. Whatever is known, it is simultaneously known that "I" know it. Who else? But which ego is it? The empirical ego of me as an individual human being, this particular carpenter, or the universal reason in which I somehow participate? The individual mind or the collective mind? From the systems theoretical point of view, the subject, whether empirical or transcendental, whether individual or collective and universal, is a construction of systemic self-reference on the level of meaning. The subject must, after all, mean something. Meaning systems are self-observing and thus distinguish themselves from the environment by a self-description which, as all information, is constructed within the system as an image or model of the system:

> ...the self-description of a social system [is] a semantic figure in which is expressed, in reduced and simplified form, the specificity of the system. The reduction and simplification is necessary because an operation of the system cannot refer to the real system as a whole; for then this operation would in turn belong to the system, thus would have to refer to itself at the same time, and so on in an infinite regress. A self-description is, so to speak, the core of an identity, an inner model or self-model of the system in the system, which is not directed at the most comprehensive possible representation, but at an adequate version of the basic operative structure. (Wilke 1991:135)

What Willke calls "an adequate version of the basic operative structure" of a meaning system is what traditionally has been called "consciousness," which was always a mystery and today has become even the "hard problem" of cognitive science. The self-referentiality of the meaning system produces particular logical and semantic problems, which we will discuss in detail below. Logically, these are the problems of "de-paradoxing," for the system cannot be reproduced within itself without contradiction. If the system observes itself as "I," who is the observer of the "I"? If the observer is also "I," we are trapped in a tautology. "I" is "I". If we de-tautologize the I, we get a paradox, since I becomes "not-I." The paradox must be de-paradoxed for system operations to continue and not be blocked by the contradiction. The system must find a way out of the tautologies and paradoxes of self-reference to enable connecting operations. This is at least a logical requirement of the autopoiesis of the system.[97]

Finally, there is a third form of self-reference insofar as the system as a whole is considered the agent. This is the place traditionally occupied by universal reason, the transcendental subject, or God. It could be called the metaphysical self-reference of the system. If you ask meaning what it is, it will answer Being, and if you ask Being what it is, it will answer meaning. At least, this is what Heidegger heard it say. The question of agency and ascription of observing to an observer is answered by reference to the system as a whole and not to any instance, such as a subject or a cognitive agent within the system. On this level of self-observation, there is no subject. The system operates by constructing meaning. The system is the actor. Heidegger will say, "language speaks," and Luhmann will repeat, "communication communicates." There is no one doing this. Neither language nor communication are in any sense of the term subjects. On the contrary, subjects, persons, cognitive agents, egos, etc. are not only particular distinctions within the system, but they know that they are not the creators of the "world" in which, as Heidegger

[97] The fact that semantic and pragmatic requirements of autopoiesis do not let themselves be blocked by purely logical problems goes, of course, without saying. No one goes through life without contradictions and no religion or worldview is logically consistent.

puts it, they are "thrown." No inner-worldly subject ever created the language it uses, the social order in which it lives, the institutions that govern its life, and the culture and religion that determines its horizon of possibilities. Let us recall that Descartes' systematic doubt and the wish to overthrow all tradition and begin anew based on certainty could not get the job done without the help of God. The system is the agent, and the system is not a person, an individual, or an empirical ego. It may, however, be thought of as a kind of super subject or as God. Indeed, this was what theology and much modern philosophy does. Kant's "transcendental ego" was the function of systemic self-reference previously occupied by God and was then replaced by pure reason and, finally, in the theory of social systems, by society. Durkheim, as is well known, referred to society as a "secular god." Since Durkheim, systemic, or metaphysical self-reference has become the historical and empirical a priori of society about which post-structuralism and post-modernism happily theorize. On this third level of self-reference, we have anonymous processes of making distinctions, constructing information, and nothing else. This type of self-reference is important in Luhmann's theory of social systems because, as we shall see, it will be the social system in the form of communication that is the primary reference for meaning and not psychical or cognitive systems, as is the case even today for philosophy of mind and cognitive science.[98]

The status of the second form of self-reference, that is, the self as a human individual or person, the ego, remains ambiguous in Luhmann's theory since, on the one hand, Luhmann defines psychic systems, that is, brains and individual human beings, as a kind of meaning system, but on the other hand, must exclude them from the social system. Luhmann

---

[98] Cognitive science, with the exception of "non-Cartesian", or "4E" cognitive science, as well as philosophy of mind largely ignore sociology and this to their own detriment. They bar themselves from exploiting the theoretical advances of sociological systems theory and are locked into biological or machine models. When they speak of "mind" they mean either brains or computers, but acknowledge nonetheless that minds must be somehow connected and that cognition is somehow distributed and collective and that there must be some kind of mind, what kind?, where?, beyond individual brains or computers. Many refer at this point to the pure consciousness of Eastern spirituality. One could get the impression that many cognitive scientists are Buddhists.

seems unable to do away with human beings as producers of meaning but cannot explain what meaning they produce. The moment they open their mouths and say something, they become socially constructed "persons" (*persona*) through which something else speaks and are no longer a-social individuals. It is difficult not to think of Hobbes' description of the state of nature in which isolated humans roamed the Earth continually at war with each other until they entered into the social contract and became citizens. We will return to this typically modern problem and its many ramifications later.

Autopoiesis: According to Luhmann, meaning is autopoiesis par excellence because the processing of meaning via the distinction between actuality and possibility is a constant selecting, relationing, and steering, which is nothing other than self-organization or autopoiesis. If one is working with a hammer, one is looking for the nails, positioning the wood, following the blueprint, etc. As soon as one thing is in the spotlight, many others are waiting just off stage. As Husserl and Heidegger well knew, meaning is the temporal processing of actuality and possibility. Luhmann reinterprets the dynamic restlessness of human knowing and consciousness in systemic terms as autopoiesis. In the operations of the meaning system, each thematic content decays in time, thus forcing further selections that run along the referential lines disclosed by the relevant horizon:

> …meaning processing constantly shapes anew the meaning-constitutive difference between actuality and potentiality. Meaning is the continual actualization of potentialities. But because meaning can be meaning only as the difference between what is actual at any moment and a horizon of possibilities, every actualization always also leads to a virtualization of the potentialities that could be connected up with it. The instability of meaning resides in the untenability of its core of actuality, the ability to restabilize is a provided by the fact that everything actual has meaning only within a horizon of possibilities indicated along with. And to have meaning means that one of the possibilities that could be connected up can and must be

> selected as the next actuality, as soon as what is actual at the moment has faded away, transpired, and given up its actuality out of its own instability. Thus, one can treat the difference between actuality and possibility in terms of temporal displacement and thereby process indications of possibility with every (new) actuality. Meaning is the unity of actualization and virtualization, of re-actualization and re-virtualization, as a self-propelling process (which can only be conditioned by systems). ... The auto-agility of meaning occurrences is autopoiesis par excellence. (Luhmann 1995:65/66)

As we shall see, the autopoiesis of the social system is communication. One communication must lead to other communications, or the autopoiesis of the social system stops, and society disappears.

Operational Closure: The fifth systems theoretical resource that Luhmann calls upon to conceptualize meaning is operational closure. Meaning becomes a fundamental concept in systems theory insofar as it is understood as the operational closure of a particular kind of system:

> Meaning always refers to meaning and never reaches out of itself to something else. Systems bound to meaning can therefore never experience or act in a manner that is free from meaning. They can never break open the reference from meaning to meaning in which they themselves are inescapably implicated. (1995:62)

If a meaning system were to experience "something" meaningless, which often happens in life, then its autopoiesis would be interrupted, and the system would disappear. We know this is a possibility from various forms of psychopathology and "crises" in life. Precisely for this reason, however, the meaning system directs its operations toward itself and makes meaning out of whatever occurs. This is what psychiatry and psychotherapy do. Everything can be interpreted as a "symptom," a sign of some psychopathology. No operation of the meaning system can produce what is meaningless; it codes everything as meaning. It has always been remarked that whenever a natural catastrophe occurs, one seeks for reasons, whether it is God's wrath, the authorities' negligence, imperfect warning systems, or irresponsible behavior. Whatever disaster could be the end or limit of the system becomes a possibility within the system. The contingency of the system is transformed into an operation of the system and absorbed into the system. The meaningless appears "as" (see the hermeneutical "as") a particular symbolism of the limit of the system. For a system of meaning, the meaningless has indeed a meaning, namely, the meaning of a limit, against which one can only push but not get beyond. As Luhmann (1995:62) puts it: "Any attempt to negate meaning on the whole would presuppose meaning, would have to occur within the world. Thus, meaning is an unnegatable category, a category without difference." This is the operational and informational closure of the meaning system.

Operational and informational closure shows what "world-complexity" means. It appears as a boundary of meaning, as "unmarked state" (Spencer Brown) of all possible world-orders, or what we have previously called the absolute complexity of environment-1, often symbolized as "chaos." But it is also the system-relative environment (environment-2), the world as it is accessed within any given meaning-boundary. This is "my" world, "our" world.[99] In this world, the world we at any time live in, specifically and structurally reduced complexity emerges. Whatever world we live in, it is always and necessarily, in many ways, a familiar world, a world we know and understand despite all the absurdities and injustices of fate. The absolute complexity, or contingency of environment-1 from which all possible worlds can emerge, raises the question of why there is something rather than nothing and the utopian hope that the world will be completely different in the future. Relative contingency is concerned with how this or that particular event or idea

---

[99] On the concept of world in general in Luhmann, see Thomas (1992). Specifically on the distinction between environment-1 and environment-2 (Thomas speaks of world-1 and world-2), see 346: "In the world (W1) of indeterminable complexity, the introduction of system/environment differences leads to the emergence of system/environment-relative worlds of (system)relative determination. ... This cosmic event of meaning unfolds meaning as determination of world (W1) by a multiplicity of differing, multidimensional, but determined and system-relative actualizations of sense.

can be integrated into the world we live in and not with the contingency of the whole world. Both forms of contingency are "dealt" with by the operational closure of the system. If God should suddenly appear and announce the end of the world, prophets would arise everywhere, declaring, "I told you so." And God or the prophets would be obliged to say something about what comes next and what we should do in this situation. Even God, at least if He wants to be understood, cannot escape the operational closure of the system; there is no way out.

To summarize, a meaning system, according to Luhmann, is an operationally and informationally closed, autopoietic, self-referential system constructed by a meaning-specific system/environment boundary whose function – like all systems – is to reduce complexity. However, for meaning systems, this is done via observations and self-observations, that is, by operations on its specific level of emergent order.

2.5 Subsystemic Differentiation or Internal Complexity of the Meaning System

Based on a systems-theoretical reception of Husserl's phenomenological analysis of meaning, Luhmann specifies how the system of meaning constructs internal or subsystemic differentiation, that is, internal complexity. Internal complexity, as we saw, determines the setpoints or goals of the system and possible states of the system, which can become responses to events in the environment influencing the attainment of these goals. When dealing with meaning systems, this raises important issues concerning the commitment of systems theory to a specific understanding of society and human existence, that is, to a particular worldview. We will ask if the systems theoretical model of society, at least in Luhmann's version, is not in fact a model of modern Western industrial society, that is, a culturally and historically limited perspective. And if the systems model of meaning turns out to be bound to a description of Western modernity, what consequences does this have for the universality claim of the theory? Could another kind of society than Western industrial society, a society which Luhmann describes as functionally differentiated, also be modeled as a system? Is meaning, therefore, only something that has emerged in the modern period of the West? What about possible forms of social order that are not adequately described as systemic? We will leave these questions unanswered for the moment and give an overview of Luhmann's attempt to theorize how meaning differentiates and specifies itself within "the" world.

For Luhmann, the "self-processing" of meaning occurs via meaning-specific *information*. According to Luhmann, information at the level of the emergent order of meaning is an event that changes states in a meaning system. The states of a system are changed according to possibilities that are predetermined in the system's structures.[100] The weather report, for example, predicts a storm. Based on this information, the school authorities declare that the school will be closed that day. The weather report is information because it makes a difference; it leads to the action of closing the school. Suppose everyone already knows the storm is coming and all preparatory measures have been taken. In that case, the report is a difference; a storm differs from sunny weather, but it makes no difference. Therefore, it is not information. For Bateson (1981:582), whom Luhmann likes to quote in this context, information is a difference (distinction, observation) that makes a difference.[101] How does information arise, and what does it do? What happens in a meaning system when information "happens"?

According to Luhmann, the structures by which system states can be transformed by information are, in the first place, meaning *dimensions*. Luhmann finds three such dimensions of meaning: There is the "factual dimension," which is structured according to the double horizon of inside and outside; there is secondly a "temporal dimension," which is structured by the double horizon of before and after; and third, there is the "social dimension," which is structured according to the double

[100] See Luhmann (1995:67ff.)

[101] We will return to the notion of information in the discussion of communication below. In general, we may assume that the unit of meaning which Luhmann regards as "symbolic generalization" is information, although it must be kept in mind that under certain conditions and at certain levels of emergent order information can be distinguished from meaning.

horizon of ego and alter ego. Each unit of meaning or information is conditioned in all three dimensions simultaneously:

> This frame of reference urges every operation to locate its intended meaning within the structure of the dimensions and their horizons. Operations must carry out determinations corresponding to the dimensions – not for the sake of the operation's own determinacy, but because otherwise they could not connect to any other operations. (1995:52)

When Luhmann speaks of "connecting to other operations" he refers to the system's autopoiesis. What do these dimensions do to enable the autopoiesis of the meaning system? Whatever we can talk about, these are things that happen either in me or outside of me. This is the "factual" dimension with its double horizon of experience and action. Experience is what happens to me; it is outside me, whereas action is what I do. Whatever is said, whether I experience it as a fact or as something that I do, it is further conditioned by the social dimension since there must always be "others" who can respond in some way, agree or disagree with what I say, in short, talk about it. And finally, whatever is said, whether it is a fact or an action, and whoever says it, ego or alter ego, takes place necessarily in time, in a sequence of before and after. We must always know what happened first, who spoke first and what was said afterward, and whose turn to speak is now. Without applying these three dimensions to information, meaning cannot be processed by the communication system. No one would know what was being talked about, how to respond, or in which way to proceed. The system's operations could not autopoietically connect to further operations. Communication would come to a halt, and the system would disappear. It is precisely to ensure that this does not happen that the three dimensions necessarily come into play; that is, they are solutions to problems that the system encounters and not structures of the world either given by God or otherwise created by some agency.

The dimensions come into play through "attribution" schemes. To reduce dangerous over-complexity and thus effectively manage the production of possibilities of references by building up internal complexity, each of these three dimensions is "schematized."[102] Schematism operates via "attribution" and "provides assistance for understanding and simplification in processing complexes with open meaning that are indispensable for preserving complex systems" (Luhmann 1995:85). The factual dimension is structured according to the scheme of external and internal attribution. According to this attribution scheme, internal and external horizons are determined to be domains of either experience or action. If information is attributed to the external horizon, it takes place within the horizon of what is perceived as outside of us. We attribute the event to the world outside. It is raining. The environment is active; I am passive. This is what *experience* means. To experience something implies that something happens in the world, and I experience it. I see, for example, the book in front of me on the table. I attribute the book to the external horizon. It is not part of me. It exists in the world and presents itself to me as soon as I open my eyes. Whatever comes in from the outside is the world of experience. On the other hand, if information is attributed to the inner horizon, then we are dealing with action. I go over and pick up the book. I do something. I may be upset by the fact that it is raining or be happy. The system is the agent; the environment is passive. For example, I close the book and take a sip of coffee. This was my decision and my action. Or if I close my eyes and imagine the book lying before me on the table, the book isn't doing this; imagining something is my action. Whether drinking coffee or imagining things, the system is the actor. The imagined book is attributed to the inner horizon; it is an image, a product of the imagination. Drinking coffee, on the other hand, is something I do, and neither the cop nor the coffee experience this or act on their own.

---

[102] The reader is inevitably reminded of Kant's notion of the disordered manifold of sense impressions which must be ordered by the schemata and categories of the mind. There are many parallels between systems theory and Kantianism since information is constructed within the system out of undifferentiated perturbations coming from the environment. Similar to Kant's unbridgeable gap between what can be known and the thing in itself, in systems theory there can be no communication, that is, flow of information between system and environment. Whether this model can be used to describe meaning systems for whom the environment only exists within the system is a question which will occupy us at length later in this book.

Second, there is the temporal dimension. The time dimension is schematized according to the attribution of either stability or variability. Without change, there would be no perception of time. Information comes and goes, stays a while, or immediately disappears. No matter what the information may be, whether experience or action, it happens in time and is ordered sequentially along a timeline. First, it started raining, and then I got upset. This becomes apparent when we consider how important narrative is in creating order of all kinds. Our world, our lives, indeed practically everything, is information ordered narratively. A story is a series of events along a timeline, and everything occurs along some timeline.

And finally, the social dimension is schematized similarly to the factual dimension, that is, by attribution, but in this case, not to causes, internal or external, but to persons. Whatever is said is said by someone. Information, such as the weather report, comes from people, speakers, and communicative actions; even if I am the only one who says it, at least I have to say it or be able to say it, or it doesn't exist. The social dimension has a troublesome ambiguity because it may be God, Reason, or Language and not any person doing the "speaking." Recalling Heidegger, "language speaks," and Luhmann himself will say that "communication communicates." As we mentioned above, this is the metaphysical self-reference of the meaning system. Does metaphysical self-reference follow the attribution schematization of the social dimension? Do we need the attribution of a speaker? Perhaps we don't when the system self-reference is to itself as a whole and not to a subject or speaker within the system. But when language speaks, the problem arises of whether anyone should respond or how the conversation should continue. As Job in the Bible well knew, one should not argue with God. This problem is solved by avoiding metaphysical self-reference and attributing information to a speaker in the social dimension, thereby accepting God's personalization and all the theological issues that come with this.

In summary, it is important to emphasize that the differentiation of the three meaning dimensions listed above and the attributions of information to either experience or action, a temporal sequence, and to different speakers are required by the fact that the system must somehow punctuate, control, and structure the otherwise unstoppable and diffuse flow of references. These three dimensions are not the only internal differentiations of the meaning system. There are also fundamental distinctions between psychic and social systems, which we will discuss immediately below, as well as the differentiation of semi-autonomous social subsystems we will deal with later. For now, let us note that meaning must be ordered by the various dimensions, schemata, and attributions to create islands of unity, identity, and information. It is within the context of this problem that Luhmann speaks of "symbols." And it is with regard to the distinction between symbol and sign that the problem of the semiotic coding of the meaning system arises, as well as justification for a non-linguistic, psychic system of meaning that is distinguished from communication.

2.6 Psychic Systems and Social Systems

Apart from the internal differentiation of meaning into the factual, social, and temporal dimensions, Luhmann proposes a further fundamental differentiation of the meaning system. This is the distinction between psychic and social systems. This distinction is central to Luhmann's theory of society as a communication system and raises the more basic question about what meaning as such is. Let us recall that for Luhmann, the substrate of all system building is a "medium," and systems emerge in this medium as "form." Our supposed primal substrate, the medium, is pure diversity, or as Luhmann puts it, "loosely coupled" elements. Form, however, happens when elements somehow become "closely coupled," when the elements are selected and related to each other in specific ways that can then be steered toward goals. Once form appears in a medium, not everything is possible. Once systems are constituted, order appears where before there was disorder. To refer to meaning as a medium in which systems arise is simply another way of speaking about the specific complexity that system construction on the level of meaning reduces. Every level of emergent order, the physical, the biological, and the semiotic, has its specific complexity. The chaos of all possible relations and references, world complexity, which is the problem that meaning systems must solve, is reduced by closely coupling otherwise loosely coupled elements by selection, relationing, and steering processes. Introducing form in a medium is nothing other than the reduction of complexity. First, this is accomplished by introducing the system/environment difference. The next step is building up internal complexity. For systems of meaning, this is done by observations, that is, by introducing distinctions within the system.

Of course, the meaning system can differentiate itself however it wishes. Saussure pointed out that systems of signs and languages are "arbitrary." A system of meaning can differentiate itself into experiences and actions, temporal sequences, and social interactions, as Luhmann proposes. Or it can differentiate itself into the holy and the profane, domains of the ritually pure and impure, realms of spirits and demons, ancestors or aliens, animals, plants, humans, and things, etc. A system of meaning can differentiate itself internally in unlimited ways, as demonstrated by the surrealist classification of animals that Jorge Luis Borges supposedly finds in a Chinese encyclopedia, which inspired Foucault's (*Words and Things*) archeology of knowledge. According to this source, which bears being cited, animals are classified as "those that belong to the Emperor, embalmed ones, those that are trained, suckling pigs, mermaids, fabulous ones, stray dogs, those included in the present classification, those that tremble as if they were mad, innumerable ones, those drawn with a very fine camelhair brush, others, those that have just broken a flower vase, and those that from a long way off look like flies." A perusal of the philosophical literature dealing with mind and current cognitive science would find a no less surprising and confusing list of mental states that supposedly make up the realm of meaning, from qualia to thoughts, perceptions, and judgments on to attitudes, beliefs, perception, memory, reasoning, and much more. As social and cultural history shows, societies have been differentiated segmentally or into a stratified hierarchy, or, as Luhmann for Western modernity proposes, in semi-autonomous functional subsystems. All these and many other possibilities offer themselves when differentiating a meaning system.

With this general caveat in mind, let us follow Luhmann's description of the internal differentiation of the system of meaning. The distinction between psychic and social systems raises the question of whether all meaning is semiotically coded or only that meaning communicated in language. And if meaning as such is semiotically coded, what basis is there for distinguishing psychic systems from communication systems? Or, if meaning is not semiotically coded, how can we talk about this supposedly non-linguistic meaning? If, as Wittgenstein said, the limits of our language are the limits of our world, and we cannot go beyond these limits, what sense does it make to speak of an extra-linguistic, purely psychic domain of meaning? For Luhmann, as we shall see, it is sufficient to ban human individual psychic systems from society. Can this strategy adequately deal with the problems arising from assigning meaning at once to consciousness and society?

Contrary to the relativistic perspective that post-structuralist deconstruction takes toward systems of meaning, the differentiation between human individuals as psychic systems and society as an all-encompassing domain of communication that Luhmann proposes does not appear as a historical and cultural accident. Luhmann claims that this distinction is necessary and generally valid for all meaning. We will have occasion to ask whether Luhmann's preference for the differentiation of psychic and social systems expresses more a preference for modern Western convictions about free individuals over against society than an imperative of systems theory itself. To approach this complicated issue, we will briefly examine Luhmann's discussion of symbols and signs.

For Luhmann, meaning arises from the formation of unity in the undifferentiated flow of references. This is the most basic operation of meaning construction. Luhmann calls this the "symbol." Symbols account for the ordering or unity of references that together constitute a unit of meaning:

> The concept symbol/symbolic is meant to denote the medium in which units are formed, the term generalization the units'

> function – to handle multiplicity operatively. In very rough outlines, it is a matter of a plurality being related to a unity and being symbolized by it. (Luhmann 1995:93)

The plurality in question here seems to be the undifferentiated flow of sense impressions. If something is to become meaningful, it cannot remain a momentary perception or unnamed impression, but many such perceptions must be grouped and generalized to become a "unity." This unity is created by a symbol, a name. Out of the flow of many individual sense impressions of a certain four-legged, barking, tail-wagging animal, there arises the symbol "dog." I see a dog. Before the formation of such units of meaning, one cannot speak of meaning at all because one has nothing "fixed," nothing lasting long enough in the mind's eye to be attributed to a horizon, whether factual, temporal, or social. "Symbolic" means that an undifferentiated multiplicity has been transformed into something meaningful; that is, it has become a distinction, a difference, a punctuation, and stability in the flow of references. Symbols are, therefore, always "generalizations" because the manifold (to use a term from Kant) is thereby given enough stability to be ordered into a system of references, a "world." Despite similarities, it must be emphasized that symbolic generalizations are not to be equated to the synthesis of an undifferentiated manifold of sensations as in Kant, but the construction of repeatability and regularity which underlies all object permanence and therefore all intentionality as consciousness of something:

> Every meaningfully grasped given must not only be fully present at a moment and thereby 'fulfill' experience or action, it must also organize self-reference, thus, ensuring that, if necessary, it can be made available again in (more or less) different kinds of situations, at other points in time, with other possible partners of social communication. This re-availability is built into concrete experience and action by symbolic generalization. (Luhmann 1995:93)

The operations of the meaning system construct redundancy, repeatability, or regularity. Symbolic generalizations thus have the function of acting as structures of expectation. "Symbolic generalizations condense the referential structure of every meaning into expectations, which indicate what a given meaning situation foresees" (Luhmann 1995:96).[103] The dog I see, for example, is a unit of meaning that allows and even imposes certain expectations upon me. I expect, for instance, that the dog will not fly away, melt into water, or turn into a table, that it is something I can feed, pet, take for walks, and much more. But with this, it becomes clear that structures of expectation have the function of restricting possibilities while at the same time opening up other possibilities and making them available for connecting operations. I know at the same time how it "goes on" with the dog and also what "does not fit." Stones, for example, are usually expected to be hard, heavy, opaque, usable as building material etc., but they are not suitable for eating, and so on. In the sense of a consistent "systems-theoretical semantics," symbolic generalizations thus reduce external complexity by constructing internal complexity in the form of symbolic generalizations as sets of expected references. Because of symbolic generalizations, not everything is possible anymore. Through symbolic generalizations, things remember themselves or refer to themselves, as it were, and remain more or less bound to their specific "meaning." Indeed, meaning is constructed by this self-reference. This is the "basal self-reference" of the meaning system, which we mentioned above as the first form of self-reference distinct from the subject-reference or the system-reference.

Because of this basal self-reference of meaning, Luhmann (1995:96) thinks he has to "replace" the concept of sign with the concept of symbolic generalization. This is a momentous decision because it lays the foundation for distinguishing between two kinds of meaning systems: psychic systems and social systems. The systems theory of meaning distances itself thereby from semiotics, which equates meaning with the sign and thus with semiotic coding. According to most semiotic theories (Saussure and Peirce), no meaning can be without a sign.

---

[103] See for a discussion of expectations Luhmann (1995:267ff).

According to Luhmann, however, meaning need not be semiotically coded. This is because signs, by definition, stand for something else, whereas meaning is related to itself through symbolic generalizations and stands, therefore, only for itself.

> The reference to the world that is immanent in all meaning prevents one from defining meaning as a *sign*. One must carefully distinguish between the structure of reference and the structure of signs. The function of a sign always requires reference to something specific, while excluding self-reference. It requires asymmetrization of a basal, recursive self-reference. In other words, there is neither a sign for the world nor a sign that indicates itself. But both of these – universality and self-reference – are indispensable properties of meaning. That is why meaning is the foundational matter: a sign must have meaning to be able fulfill its function, but meaning is not a sign. Meaning forms the context in which all signs are determined, it is the *conditio sine qua non* of their asymmetrization. But taken as a sign, meaning would be able to stand only as a sign for itself, that is, as a sign for the non-fulfillment of the sign's function. (Luhmann 1995:71)

If the function of the sign is always to refer to another, to that which the sign signifies, that of which it is a sign, and on the contrary, meaning can only refer to itself, then it would seem that meaning cannot be a sign, or constituted by signs. This implies that the semiotic coding of meaning is something derivative from a primary form of meaning. But what, we may ask, is this primary form of meaning if not something we can talk about, something that is referenced by signs and thus semiotically coded? The attempt to distinguish systems theory from semiotics is counter-intuitive since it is obvious that signs also refer to themselves and that language can talk about itself. After all, Luhmann himself is talking at length about it. Meaning it would seem, can and must have a sign structure; otherwise, we couldn't be having this discussion.

In his later work, Luhmann (1993:61-62) proposed a difference-theoretic or form-analytic approach to meaning, which allows the symbol to be understood as a sign:

> In cases where the sign itself designates its own function of unifying what is separated, one could speak of symbols. A symbol would then be the self-signification of a sign. Therefore, symbolic signs are not simply guideposts that point to something else. They are not just carriers of a signifying reference and, therefore, not only materializations of the signifier. Rather they also contain a hint of the function they fulfill, that is, a hint of the very unity-constituting meaning of the sign.

Following this reasoning, there is no fundamental difference between symbols and signs. No non-linguistic or pre-linguistic symbols could be available as carriers of meaning before or outside of semiotic coding. Both symbols and signs must therefore be understood as necessarily semiotic. If they are to be distinguished at all, both fulfill the criteria of universality and self-reference that characterize meaning. Contrary to his earlier view, Luhmann grounds this possibility in the fact that the "form" (Spencer Brown) of meaning and sign are the same.[104] Both sign and meaning are the unity of a fundamental, non-hierarchical distinction. Meaning, as we saw, is understood as the unity of the difference between actuality and potentiality since every reference refers necessarily to other possible references, and the sign is the unity of the difference between signifier and signified.[105] Saussure (1967) expressed this unity between signifier and signified as an inseparable

---

[104] On Luhmann's concept of form, see (1993:49): "The distinction itself, insofar as it is distinguished from that which is distinguished by it, is then the form." Luhmann is aware that here he goes beyond what can be found in Spencer Brown.

[105] See Luhmann (1993:63): "The concept of the two-sided form, however, better expresses that the inside of the form, namely the respectively actualized sense, only makes sense with regard to the possibility of actualizing other possibilities, and that this presupposes dynamic systems consisting of operations (events). Meaning is consequently form as boundary, which is always co-observed, but can never be exceeded operatively, since every operation remains on the inside of form, namely actualizes meaning. We thus find in the form of meaning exactly the same state of affairs that the analysis of the form of signs has revealed: The operations always remain on the inside of the form. ... The operations which use the form of the sign always stay on the side of the signifier. But they need, to be able to mark this side, another side. This other side is always present as an unmarked state."

bond. So inseparable is this bond that there is no signified without the signifier, which implies that the order of things, which was supposed to lie outside the order of signs and give meaning to signs, is merged with the order of signs. Meaning cannot be derived from things that are somehow outside signification but consists in the relations of signs among themselves, that is, the system of references within language. In the same way, meaning for systems theory is the system of references constituting a world. Here, Luhmann, just like Saussure, wants to overcome the long outdated "realist" conception of the sign, which assigns the source of meaning to the thing outside of language that is then supposed to be signified by language.

> An operationally closed system using language cannot relate to the environment at the level of its operations. In this respect, there is no referent. (Luhmann 1993:51)

If one follows realism and derives the meaning of signs (semantics) from the things they refer to, then one runs the risk of reifying the sign as isolated from its meaning-constitutive references to other signs (syntax) and to the respective uses of signs (pragmatics). Semiotically and in terms of philosophy of language, such a concept of the sign is highly questionable because the "suppression of the referent," i.e., the subsuming of extra-linguistic things to language, is precisely what modern semiotics does. According to Saussure (1967), a sign is an inseparable unity of signifier and signified, whereby the signified, i.e., the thing signified, cannot be a real thing outside the differential structure of language. Signs are said to be "arbitrary" and "unmotivated" precisely because they do not read their meaning from things. But they are not arbitrary because their meaning arises from the differential relations of signs to each other. Saussure is not alone. For Peirce (1983), the object, or referent, is not something outside the sign or something other than the sign but one of the three constitutive relations of any sign: object, representamen, and interpretant; all three together make up the sign. The object belongs essentially to the sign; it is internal to the sign and cannot lie around somewhere outside it. From a systems theoretical perspective, this fundamental assumption of semiotics is accounted for by the fact that the environment of a meaning system is within the system and that the system is operationally closed.

If there can be no meaning without signs or language, then it becomes difficult for Luhmann to maintain that there are two kinds of meaning systems: psychic and social systems. If one overlooks the fundamental connection between sign and meaning, then the consequences for system-theoretical semiotics are problematic. Meaning, according to this view, cannot be equated with language since language consists of signs that are only subsequently connected with symbolic generalizations that, according to Luhmann, are already developed "in concrete dealings with objects and events" (1993:90). The distinction between "symbolic generalizations" on the one hand, which Luhmann wishes to define as the basic units of meaning, and which can be ascribed to psychic systems, and signs on the other hand, which Luhmann ascribes to communication and thus social systems grounds the distinction between psychic systems and social systems. But if there is no difference between the meaning that psychic systems construct and the meaning that social systems construct since both are semiotically coded, how are we to understand this fundamental differentiation of meaning into psychic systems and social systems? If there are psychic systems without language, signs, and communication that also construct meaning, how do they do this? What kind of meaning are we talking about, assuming, of course, that we can talk about it at all.

We probably have to think of visual perception here since Luhmann often refers to the distinction between form and medium from Heider's psychology of perception (e.g., in Luhmann 1990b:53ff.) when he wants to give an example of non-linguistically constituted meaning.[106] Thus, meaning can be distinguished from communication, which is, after all, language-dependent. Language and communication are, therefore, not the forms par excellence in which meaning is constructed. According

[106] See Luhmann (19995:412): "Perception is a less demanding form of acquiring information than communication. It makes possible information that does not depend on being selected and communicated as such." But, of course, it could be communicated because it is semiotically coded. Whether it is less demanding not to say what one sees or is thinking than just blurting it out is an open question.

to Luhmann, communication becomes a particular form of processing meaning among other forms. This assumption does not change in his later difference-theoretical semiotics.[107] Despite his ambivalence on the nature of the sign, Luhmann consequently adheres to the distinction between psychic and social systems, that is, to a fundamental difference between individuals and society. But we much ask if this typically modern Western view of free, autonomous individuals existing in a Hobbesian "state of nature" somehow before and outside of society is compatible with a systems theory of meaning or whether we are dealing with ideological commitments to Western modernity that systems theory does not need. Must the meaning system differentiate itself into individuals over against society?

Luhmann's insistence on the typically modern distinction between individuals and society leads him to assert that semiotics and hermeneutics cannot provide an adequate theory of meaning. Signs have to be interpreted, i.e., they require a hermeneutic understanding. But hermeneutic interpretation presupposes that a unit of meaning is embedded in another horizon. This is the effect of the social and temporal schematization of meaning. It appears as consensus or disagreement about the past, present, and future. Gadamer (1989) referred to this basic foundation of hermeneutics as "fusion of horizons" (*Horizontvechmelzung*). Because other people and other times have their own horizons of meaning, this difference must be bridged in present-day understanding of others and of tradition. This fusion of horizons is what makes up hermeneutical understanding. Since the social horizon is not available to isolated psychic systems, Luhmann must claim that the schematization of meaning in a social horizon is unnecessary for meaning. This implies that there is information that supposedly does not need to be understood. What is this information? According to Luhmann, these are the mental states of psychic systems. Psychic systems process meaning in a purely inner, private, signless flow of perceptions and thoughts.

[107] See Luhmann (1993:64): "The sign-form is *one* (emphasis AB/DK) among several possibilities to translate the paradox of meaning, the indeterminacy of the determinate, into a distinction and thus to unfold it." What are the other possibilities if we cannot designate them? Here again, the assumption seems to be that meaning without signs is possible.

Meaning is constituted unmediated and directly rather than mediated through sign use. This idea is evident in Luhmann's (1993:53) "substitution" of Peirce's concept of "thirdness" for the concept of "observation." According to Peirce, thirdness underlies the interpretant; it is the sign's meaning or how it is finally interpreted. For Peirce, the interpretant denotes the application of a rule or convention that governs the associations a sign might bring with it. Observing, however, is not an operation that applies a rule or convention. Herein lies, we may assume, the reason why pragmatics, the study of how signs are used according to rules, is neglected in systems theory.

This foreshortening of the concept of sign in Luhmann's systems theory has serious consequences in that it allows Luhmann to postulate two kinds of systems. Luhmann's decision to take Husserl's phenomenology with its attachment to the ego/consciousness as a starting point inherits Husserl's solipsism and leads his theory of meaning to distinguish between systems of consciousness, which process meaning in the form of a purely internal, pre-communicative (and consequently pre-semiotic) flows of thought and perception, and social systems, which process meaning in the form of communications.[108] According to Luhmann (1995:98), psychic systems are distinguished from social systems in that meaning is reproduced in different forms in each case:

> Meaning can insert itself into a sequence that is attached to bodily feelings, then it appears as consciousness. But meaning can also insert itself into a sequence involving the understanding of others and then appear as communication.[109]

---

[108] It is interesting to note that this assumption is foundational for cognitive science in its various Cartesian forms.

[109] Luhmann insists that systems of consciousness and systems of communication must be strictly distinguished. Indeed, as systems, they are constituted by mutual exclusion, that is, by the system/environment difference. Thus Luhmann can say (1990b:32): "Systems of consciousness and systems of communication exist separately." Even more precisely (45,46): "Systems of consciousness are operationally closed, self-dynamic, i.e., restless, mercurial systems of the reproduction of their own thoughts by their own thoughts. ... ...a communication system is a system coupled to consciousness, irritated by consciousness, but which can determine its own operations only by its own structures and its own structures only by its own operations."

There are thus two kinds of meaning systems, the psychic and the social, with the psychic system consisting of a sequence of mental states that are not semiotically coded and the social system consisting of a series of linguistic communications.

Meaning can thus be understood independently of language, communication, and action, as in the Western metaphysics of the subject from Descartes to Husserl. In this view, meaning theory remains committed to Western modernity. However, one of Luhmann's theoretical innovations is to strictly separate the two types of systems and to emphasize, completely contra-intuitively, that society does not consist of human beings but of communications. Human beings are banned from society for the first time in Western thought.[110] Nonetheless, there are no communications without mental states, no social system without consciousness. Luhmann conceptualizes the relation between individual and society according to the scheme of system and environment and appeals to a biological metaphor.

> This is precisely how the chemical system of cells is environment for the brain and how a person's consciousness is environment for the social system. No decomposition of neurophysiological processes could ever reach individual cells as its ultimate elements and no decomposition of social process could ever arrive at consciousness. (1995:179)

Is this distinction between two forms of meaning systems tenable? Is meaning conceivable without semiotic organization? Can a purely private, non-semiotically coded stream of consciousness be conceived as a meaning system? Is purely "private" observation as the handling of

---

[110] Thus, according to Luhmann (1990b:30-1) in an extreme formulation, "There are thus no 'conscious communications', just as there is no 'communicative thinking' (sensing, perceiving). Or, to put it another way, human beings cannot communicate...." Luhmann (1995:210) sees the main difference with tradition in that for "the humanist tradition human beings stand within the social order and not outside it. ... If one views human beings as part of the environment of society (instead of as part of society itself), this changes the premises of all the traditional questions, including those of classical humanism" (212). With these statements Luhmann clearly takes a "post-humanist" position.

differences possible? Or can differences only be handled based on semiotic organization, which implies communication? Is the specific system/environment difference that constitutes social systems a difference between consciousness and communication, or would it not have to be understood as a difference between external and internal communicative contingency? Even if I don't talk to others, I am constantly talking to myself, and for this very reason, I am able to talk to others and they to me. Even if my feelings and thoughts are my own, they are so constituted that I could talk about them if I wanted. Is it not so that all systems of meaning are semiotic systems whose elements are selected, related, and controlled by semiotic organization? Must we not conclude that insofar as the human being or consciousness functions as an element of the meaning system, it functions only, as Peirce said, as a sign or as Luhmann himself will say, as a socially constructed "person?"

These questions are at the basis of our critique of Luhmann's theory. Therefore, in the following discussion, we will attempt to question the distinction between the two types of meaning systems, and this must be emphasized, without denying the existence of individuals or of consciousness or, on the other hand, advocating any form of collectivism in opposition to individualism. Instead, it is a question of the importance of communication for the constitution of meaning and, consequently, for the construction of individuality, consciousness, and society in communication. Contrary to Luhmann, we ask: Can meaning be adequately described without reference to signs, language, communication, and understanding? Does not all information, insofar as it is conceived as distinction and designation, necessarily imply coding, and does not coding at least semiotic coding – in turn, necessarily imply communication and understanding? Is not a system of meaning necessarily semiotically organized?

If it turns out that individual consciousness is not even possible outside the framework of the collectively and socially constituted communication community, there would be no evidence in systems theory to postulate two kinds of meaning systems. The two domains would coincide, meaning that meaning could not be considered pre-social and pre-communicative. In other words, if systems of consciousness could not be distinguished from social systems, at least at the level of meaning constitution, there would be no psychogenesis of meaning independent of sociogenesis, no individual without society, no brain without other brains, and constitutive relations to the world.[111] Consciousness, whatever it might be, becomes much less important for understanding meaning than it currently is in most discussions of the philosophy of mind and cognitive science. If it turns out that cognition – also cognition about one's own "bodily feelings" – is only possible if it is semiotically mediated and intersubjectively accessible and controllable, systems of meaning cannot be divided into two species – one psychic and one social – which would then somehow be "adapted," or "structurally coupled" to each other in a kind of symbiosis.[112]

The upshot of this discussion is that meaning systems can handle differences only under the conditions of communication. Communication can differentiate its ascription of agency however it wishes. The meaning system can construct any kind of subjects or identities that it wants. There is no necessity that we must concern ourselves with individuals and society, with psychic and social actors. At the level of emergent order that we call meaning, we are never dealing with single individuals and their private, purely internal cognitive states or observational operations, but always, directly or indirectly, with a system of communication as an intersubjective whole within which such distinctions as those between individual and group are used for specific purposes in specific historical circumstances.

Thus, as a counter-thesis to Luhmann, we claim that meaning must be conceived of at the communicative level. Communication must be understood as a foundational condition of meaning, not vice versa. This does not prejudge whether one can continue to speak of social

---

[111] According to 4E cognitive science, mind is embodied, embedded, enacted, and extended into the environment. See Rowlands (2010).

[112] Luhmann doesn't speak of symbiosis, but of "interpenetration" (1995:213ff.) to describe the relation between psychic systems and social systems. We will return to the term "interpenetration," which Luhmann following Parsons takes to refer to the relationship between humans and society below.

systems and mental systems as particular internal differentiations within communication systems. The thesis claims only that systems of meaning are necessarily semiotically coded systems of communication, no matter how they may be further differentiated and specified.

To make this thesis plausible against Luhmann's view, we have to resort to the linguistic turn in the philosophy of language and epistemology. In short, the meaning-constitutive role of communication is shown by the fact that internal contingency, i.e., the non-arbitrariness of structures of expectation, is not conceivable outside or before intersubjective corrigibility, i.e., through communication. No distinction, observation, designation, or symbolic generalization is conceivable as a pre-communicative product of a not-yet-social system. Thus, in place of the transcendental subject, there are not two structurally coupled kinds of meaning systems but the linguistically constituted community of communication.

We assume that units of meaning reduce contingency by creating redundancy. According to Wittgenstein, who is close to Peirce's pragmatic semiotics in this respect, symbolic generalizations are to be understood as "rules" (Peirce speaks of conventions) because a rule cannot be followed only once. We quote this passage (Wittgenstein 1953:§199) again:

> It cannot be only one time that only one person has followed a rule. It cannot have been only once that a communication was made, a command given, or understood, etc. – To follow a rule, to make a communication, to give a command, to play a game of chess are customs (usages, institutions). To understand a sentence is to understand a language. To understand a language is to master a technique.

According to Bateson, a unit of meaning is constituted by communication because only information can be perceived. Information is a difference that makes a difference. Such a difference must be communicative because it could not make differences without selection, i.e., encoding and communication, without which it could have no state-changing effect for any system. A difference that makes no difference is not information. For example, if we know what "smoke" is, we expect that there is fire where there is smoke, and we can respond accordingly. Not everything is possible in a coded world where rules of association and reference apply. The complexity of the world is reduced to a certain regularity. For Bateson, as for Wittgenstein, the reference/horizon relation constituting all meaning is constituted by redundancy. Unlike Luhmann, however, for both Bateson and Wittgenstein, there is redundancy only through linguistic and semiotic organization. As Bateson (1972:421-2) says:

> I would argue, however, that the concept 'redundancy' is at least a partial synonym of 'meaning.' As I see it, if the receiver can guess at missing parts of the message, then those parts which are received must, in fact, carry a meaning which refers to the missing parts and is information about those parts. ... In sum, 'redundancy' and 'meaning' become synonymous whenever both words are applied to the same universe of discourse.

How else do we imagine the possibility of reduction of contingency if not through language? How could the psychic system organize, condition, and relate many thoughts, feelings, imaginings, intuitions, sensory and bodily perceptions, fantasy images, moods, desires, impulses, etc., if not through a semiotic code? Without semiotic coding, that is, communication, we have only autism. To disrupt the autopoiesis of a psychic system, the organism doesn't need to die. All that needs to happen is that the operations of the system must become disorganized, and the system will dissolve into madness. The specific contingency of consciousness must be reduced and organized by some principle. This principle can be only language, for no other principles of meaningful organization remain. As Wittgenstein showed, there can be no private language; the idea of a self-contained, isolated psychic system is absurd. Below the language boundary, no consciousness is possible. And above the language boundary, beyond language, no isolated, purely private, a-semiotic consciousness is possible. Even the mystics only know about their silence because they can talk about it, and this is when they say it is ineffable. The idea of a pre-communicative, pre-linguistic system of consciousness is neither empirically nor logically possible. If such a system of pure, non-communicative consciousness were empirically present, it would have to be able to make itself known, for which it would then have to make use of language and communication. If it could not do this, it would dissolve in the senselessness of an arbitrary jumble of states of consciousness. Finally, as Habermas points out, meaningful complexity reduction need not be explained on the basis of the model of intentional consciousness:

> If we now abandon the basic consciousness-philosophical concepts in which Husserl treats the lifeworld problem, we can think of the lifeworld as represented by a culturally transmitted and linguistically organized stock of interpretive patterns. Then the talk of a referential context that connects the situational components with each other and the situation with the life-world no longer needs to be explained within the framework of a phenomenology and psychology of perception. (Habermas 1981b:189-90)

Let it be said again, we do not wish to advocate collectivism of any kind by rejecting Luhmann's' notion of individual psychic systems. The dichotomy of individualism and collectivism is a typical structure of modern Western culture and society. Not having one does not imply you are automatically stuck with the other. The solution is to have neither. Our critical remarks on Luhmann's theory of meaning are not intended to assert that there are no individuals or that subjectivity in the sense of a private, emotional, and thoughtful interiority is illusory. We merely advocate that individuals and society be understood from an adequate theory of meaning. Such a theory must distinguish levels of emergent order. It must distinguish between systems of meaning, on the one hand, and mechanical and biological systems, on the other. It must be able to account for the difference between genetic and semiotic coding theoretically. We claim that there is no other form of organization for meaning systems than semiotic coding.

If talk of mental and social systems is to have any meaning at all, and even if only to serve as a fruitful guiding idea for different research approaches (e.g., in cognitive science, psychology, and sociology), then both types of systems should be considered as constructs of the one semiotically constituted meaning system. Accordingly, we emphasize that distinctions such as those between the psyche and society do indeed have meaning. They are common and useful distinctions in everyday understanding and as demarcations between scientific disciplines. It is, interestingly, the everyday understanding that Luhmann draws on when he wants to make his distinction between thinking and talking plausible: we say something and think something else at the same time; we cannot perceive another person's perceptions directly; we think associatively and only in fragments of words and sentences and not according to the regular flow of linguistic communication, etc.

All this is just as true as it is irrelevant. Such distinctions have been conceived for purposes other than to ground a theory of meaning. They are far too undifferentiated or, as Luhmann says, have too little "resolving power" to serve as the foundations of a solid theory of meaning. They presuppose meaning as already given. Accordingly, they cannot be applied without fundamental reinterpretations at the level of a theory.

With these arguments in the background, we now turn to the specifically social form of meaning that Luhmann calls "society." So far, we have critically examined the form of meaning-processing that Luhmann attributes to psychic systems. We have pursued the question of whether the distinction between two kinds of meaning systems can itself have meaning. The argument was directed against the idea of an a-semiotic, pre-communicative meaning and thus against the idea of two kinds of meaning systems. It will be seen in the discussion below that even in the specifically social realm, the preliminary decisions of Luhmann's theory of meaning lead to problems. If psychic and social systems are so different, it has to be explained how they relate to each other. How do they condition each other? We begin by asking how social systems can arise from psychic systems.

2.7 The Social System

We saw that one of the first major internal differentiations of the meaning system was the introduction of three fundamental dimensions of all meaning: factual, temporal, and social. Each dimension is "schematized" into two mutually dependent horizons. In the social dimension, all meaning is attributed to either ego or alter-ego. The social dimension conditions information such that it flows between ego and alter ego. This is communication. This is also, for Luhmann, the specific social system in distinction to individual, non-communicative psychic systems. We have argued at length above against the idea that there can be individual, non-communicative psychic systems. We need not repeat these arguments here. Nonetheless, since, for Luhmann, the emergence of meaning is not equivalent to the emergence of society, the theory of social systems must explain how communication and thus society come into being on the basis of non-communicative psychic systems. He attempts to answer this question in two ways. First, he refers to Parsons' sociological theory of double contingency. Secondly, he proposes a systems theory explanation of how individuals and society relate to each other in terms of what he calls "interpenetration." We will briefly look at both these attempts to explain society beginning from individuals even though, as argued above, we view this entire enterprise as misguided and unnecessary.

To explain the specific social processing of meaning, Luhmann first introduces the model of double contingency.[113] Double contingency is intended to explain how the social as a realm of commonly recognized, i.e., stable, and repeatable structures of expectation emerges

[113] In the original version quoted by Luhmann (1995:103ff.) from Parsons and Shils' *Towards a General Theory of Action*, "There is a double contingency inherent in interaction. On the one hand, ego's gratifications are contingent on his selection among available alternatives. But in turn, alter's reaction will be contingent on ego's selection and will result from a complementary selection on alter's part. Because of this double contingency, communication, which is the preoccupation of cultural patterns, could not exist without both generalization from the particularity of the specific situations (which are never identical for ego and alter) and stability of meaning which can only be assured by 'conventions' observed by both parties."

from the realm of isolated, pre-communicatively constituted systems of consciousness. The following considerations are concerned with the problem of the emergence of the social as a realm in itself in contrast to the realm of individual, isolated psychic systems. The introduction of Parson's theory of double contingency serves to synthesize three theoretical traditions: Parson's explanation of the emergence of society, Husserl's phenomenological analysis of intentional consciousness, and systems theory's requirement that a system must be founded on the system/environment difference.

The first step in this synthesis is to extend the concept of contingency to its "original modal theoretical version."

> Something is contingent insofar as it is neither necessary nor impossible; it is just what it is (or was or will be), though it could also be otherwise. The concept thus describes something given (experienced, expected, remembered, fantasized) in the light of its possibly being otherwise; it designates objects within the horizon of possible variations. (Luhmann 1995:106).

The idea of contingency is to be understood in terms of Husserl's notion of intentional consciousness of an object against a horizon of possible references or variations. If everything that appears, appears "as" this or that (the hermeneutical "as"), then it can appear as something else from a different perspective or in a different context. All meaning is inherently contingent. According to Luhmann, double contingency has two consequences:

> It enables the differentiation of a particular *world dimension* for socially distinct meaning perspectives (the social dimension), and it enables the differentiation of particular *action systems*, namely, social systems. (106)

The concept of contingency, therefore, connects Parson's double-contingency model with Husserl's idea of intentionality as the consciousness of an object against a perspectival horizon. These two together form the background for explaining a unique system/environment difference that constitutes social systems as specific forms of meaning systems. To what extent is this strategy successful?

Let us base the following discussion on a simple example. Two children are playing next to each other on a playground. Both have a ball in their hands. Initially, they both, each separately, engage in an indeterminate behavioral variation. The ball is pushed back and forth arbitrarily with the hands, then with the feet; it is thrown into the air and dropped on the ground or caught with the hands. There is dancing around the ball, jumping over it, sitting on it, etc. During all this, the children do not follow any behavior rules nor interact with each other. There is no question of a social system yet. Then, one child throws the ball to the other. The second child also selects this behavior; he throws the ball back to the first child. This behavior is repeated, resulting in mutual behavioral selection, which is then stabilized; it can be continued or stopped and then taken up again. A structure is formed, distinguished from the arbitrary variations of behavior that occurred before. One could say that the social system "ball game" has come into being. The structure called the ball game, remains even when the children go home for lunch because when they meet again at the playground afterward, they can resume the ball game.

This model describes the situation of double contingency in that 1) the behavior of each child is contingent, free, and undetermined, 2) both children know this of each other, and 3) on the basis of interaction, both become dependent or conditioned by each other's behavior. They have mutual expectations as to what action they should each perform. Each child knows that the other knows that it knows what is expected. And they honor these expectations and accept the "obligation" of acting according to what is expected. Without mutual acceptance of the rules of the game, the game would not exist. Without mutual acceptance of rules governing social behavior, as Hobbes well knew, society could not exist. Thus, it can be said that a social interaction or a social system emerges autocatalytically via variation, selection, stabilization, and structure formation under the conditions of double contingency. But what does this have to do with meaning?

It is important to note that double contingency presupposes the recognition of the other as alter ego. The ball game, which is a simple model of society, arises not as a result of solving the problem of double contingency but on the basis of mutual recognition, which makes interaction and thus double contingency possible. Why play ball with something that cannot understand what I mean to do by throwing the ball to it? Luhmann refers to this foundation of the social as "respect" (*Achtung*), and he asserts quite plausibly that without recognition of the other as alter ego, and vice versa, there can be no social interaction, no communication, and no society. Even when children play with their toys, they speak to and relate to them as though they understand what is expected of them and can respond appropriately. One can only have social expectations with regard to others who are recognized to be capable of communication. Thus Luhmann (1995:121-2):

> Ego experiences alter as alter ego. But along with the *nonidentity of perspectives*, ego also experiences the *identity of this experience* on *both* sides. The situation is indeterminable, unstable, and unacceptable for both the participants. In *this* experience the perspectives *converge*, and that makes it possible to suppose an interest in negating this negativity, an interest in determination.

Just as in Hobbes' state of nature wherein all were at war with all, and there was no common acceptance of any expectations or rules, all parties realize that this chaotic and stressful situation must be negated and in some way made determinate, that is, meaningful. Meaning, therefore, is not only about mutually accepting rules of behavior or honoring expectations. It is also not reducible to Husserl's intentional thematization of some content or information against a horizon of possible further references. What prevents such a solipsistic consciousness from becoming arbitrary and chaotic? Let us recall that the word "conscious" comes from Latin and means "knowing together." For Luhmann, meaning must arise as a social system to the extent that information and mutual recognition become a self-referential, operationally closed system. This is a system of communication that excludes all that is not communication. This is society.

The moment an isolated psychic system recognizes another as like itself, it is no longer an isolated psychic system, but is transformed into something else. Thus, there is either a plurality of isolated conscious systems, which then cannot be conceived as a social unity and therefore also cannot stand in a situation of double contingency, or there is a multiplicity of communicative agents which already are the social system; they belong to society and not merely to the environment of society. Accordingly, when the children were playing side by side before they had started their ball game, it must be assumed that this behavior was already social. It took place on a playground conditioned by many relations and references and is not conceivable as an indeterminate and unstructured behavior in the sense of a pre-social substrate of society. The children are already "socialized" to play with balls before they appear on the playground. The situation of double contingency, therefore, does not model the emergence of communication. Were the children not already communicative, we would be dealing with psycho-social pathologies such as autism. Thus, the social does not emerge from a non-social domain of some purely psychic meaning, just as Husserl can't escape solipsism once he begins with a lonely Cartesian ego. Even Descartes had to call upon God to save him from this untenable situation. In conclusion, we must admit that the theory of double contingency must presuppose that which it is supposed to explain.

If one nevertheless wants to try to use the model of double contingency as a bridge between pre-socially conceived individuals and the social, one must think of it behavioristically, i.e., as a model of mere behavioral coordination or, according to Maturana, as "structural coupling" among systems which are environment for each other. Indeed, like Maturana, Luhmann seems to want to ground the social on the non-meaningful communicatively conceived feedback loop of mutual observation, which the behavioristically interpreted model of double contingency depicts. The substructure presupposed in the theorem of double contingency presupposes highly complex meaning-using systems that are non-transparent and non-calculable to each other:

> ...they do not understand each other ... They concentrate on

> what they can observe as input and output in the other as a system in an environment and learn self-referentially in their own observer perspective. They can try to influence what they observe by their own action and can learn further from the feedback. In this way, an emergent order can arise that is *conditioned* by the complexity of the systems that make it possible, *but that does not depend on this complexity's being calculated or controlled.* We call this emergent order a social system. (Luhmann 1995:110)

As discussed above, the fact that "highly complex meaning-using systems" appear to each other in the social dimension of ego and alter ego means that the problem of the emergence of the social is already solved in the presupposition. The behaviorist alternative, however, tries to avoid ascription of subjective states by observing the other from the outside. Therefore, it cannot explain how the alter becomes an alter ego and, thus, how society could emerge from the interaction of ego and alter ego. If two systems observe each other as system-in-an-environment, for example, we observe the frog in the pond, and the frog is watching us too, this does not mean that we are dealing with expectations of expectations, i.e., with meaning or with society. We could take the frog as an alter ego, and then we would have a social situation as we often do with pets and children with their toys. Mutual observation and behavioral coordination, which happens among organisms in a niche, do not have anything to do with double contingency or with the construction of commonly recognized expectational structures. Behavioral coordination as adaptation occurs between any organism and its environment continuously, i.e., as long as the autopoiesis of the organism is maintained. For this purpose, meaning is not necessary. The question arises: Can meaning systems observe each other as system-in-an-environment so that communication and social interaction come into being only based on such mutual observation? Or must not communicatively constituted meaning be regarded as always already conditioning any observation?

According to the behaviorist view of double contingency, the social is to be built up only by the interaction of independent organisms with their own areas of experience. Two "black boxes" react to each other until behavioral coordination occurs, which Maturana considers as a structural coupling or the emergence of a "consensual domain." This model can be adopted from phenomenology in that it begins with a non-socially conceived individual consciousness as a putative system of meaning. But this only exchanges the problems of biological reductionism with the problems of idealism. It leads to exactly the same problem that Husserl was unable to solve, namely the problem of solipsism. How is it possible to move from a purely private experience to the experience of the world shared with others in language and communication?

Thus, if it does not fail because of idealism, the attempt to derive the social as an independent system of meaning from the model of double contingency leads back to biological reductionism. One has little reason to speak of systems of meaning as a separate level of emergent order beyond life. Only if the situation of double contingency is already conceived as a meaning system and already as communicative can it serve as a model to explain the emergence of expectations and coordinated action in society, but then we do not need it. The explanation is already in the presupposition. The recognition of the other and the necessity of understanding is presupposed so that the two "black boxes" could even come up with the idea of talking to each other. Communication is already there as a condition of the possibility of interaction. Communication cannot be deduced or explained from isolated conscious agents.

We can now turn to Luhmann's second strategy to get the social out of isolated psychic systems. This is the idea of "interpenetration." Luhmann (1995:213ff.) speaks of "interpenetration" to designate the relationship between human individuals and society, i.e., between psychic systems and communication systems. The term is meant to answer the question about the conditions of the possibility of double contingency and, thus, society.

> Interpenetration is not a general relation between system and environment but an intersystem relation between systems

> that are environments for each other. … the concept of interpenetration indicates a very specific situation, which must be distinguished above all from input/output relations. We speak of "penetration" if a system makes its own complexity (and with it indeterminacy, contingency, and the pressure to select) available for constructing another system. Precisely in this sense social systems presuppose "life." Accordingly, interpenetration exists when this occurs reciprocally, that is, when both systems enable each other by introducing their own already-constituted complexity into each other. (Luhmann 1995:213)

This remarkable statement poses several questions. How can a system be related to something that is neither itself (self-reference) nor the system's environment? What does it mean to say there is a relation between systems that are environment for each other but that this is not a relation between a system and its environment? An autopoietic, self-referential, operationally and informationally closed system cannot communicate with the environment. It can only be perturbed by changes in the environment and must construct information out of these perturbations according to its organization. If it does this successfully, it can continue its autopoiesis and, for an observer, appears to be adapted to the environment. How can such a system "introduce its already constituted complexity (internal complexity) into another system"? What could it possibly mean, based on general systems theory, for a system to "introduce" internal systemic complexity into another system? Furthermore, how can this happen between systems reciprocally? In other words, how can a system accept, take in, and process the internal complexity of another system that is environment for it? And finally, if such a thing could happen, are we talking about emergence, or are we talking about self-organization and internal differentiation?

These questions lie at the heart of Luhmann's theory of social systems since the concept of interpenetration is called upon whenever it becomes necessary to explain how information can be exchanged between closed systems in society. We will see below that the social system differentiates itself into many different kinds of subsystems, large functional subsystems such as business, science, education, law, etc. Then, within these functional subsystems, all sorts of organizations arise, which are also systems. As general systems theory requires, all these subsystems are modeled as autopoietic, self-referential, and informationally and operationally closed systems. And all are merely environment for each other. If there can be no communication between system and environment, how can all the different systems of which society is composed share information and cooperate to build a unified society? The answer that Luhmann gives is interpenetration. Therefore, we shall take the opportunity that the problem of the relationship between individual psychic systems and society offers to examine the concept of interpenetration more closely.

Let us take an example to illustrate the conceptual problems that the concept of interpenetration brings with it. A frog lives in a pond that is full of other organisms. The frog is adapted to this niche, or more correctly, structurally coupled to various other organisms and natural conditions in this niche. For example, the frog is structurally coupled to the pond's water and the insects it eats. However, it is not structurally coupled to the concert being performed in the park nearby. In this situation, is there any "intersystem relationship" that is not a relation between system and environment and thus a relation of structural coupling? The answer must be no. All systems in this niche, so long it can be considered a balanced ecosystem, are adapted to each other and structurally coupled. We can call this the "normal" situation. Therefore, it is not surprising that Luhmann must emphasize that the concept of interpenetration designates a "very specific situation," a situation we do not normally encounter.

What is "very specific" about the situation of interpenetration is that we have to do with systems that somehow "introduce" their internal complexity into each other, even though there can be no communication between the system and the environment. The internal complexity of a system can be understood as the elements that are selected and related to each other and thus constructed to perform the operations that the system needs to maintain its autopoiesis. If the system is an autonomous, self-regulating learning system, it processes environmental perturbations according to its internal complexity to produce information that steers its operations. We must assume that the internal complexity that is being made available to other systems in interpenetration is information since, otherwise, the systems would have to eat each other to access their internal complexity. One could suppose that the fly could make its internal complexity available to the frog for its self-construction, but this means the frog eats the fly. Only if we are talking about information can interpenetration mean that systems can somehow "introduce" internal complexity into each other. The exchange of information, however, is communication.

The situation of communication is indeed a "very specific" situation. We do not observe this situation when describing the frog in the pond or any physical or biological phenomena. We observe communication only within communication and not as something that happens between communication systems and their environment. If we suppose that systems communicate with each other, they can do this only because they are no longer autonomous systems but elements of an encompassing communication system. The moment Husserl's solipsistic intentional consciousness opens its mouth and says something, it disappears and becomes a construction of the social system, a necessary ascription of the communicative act. Communication communicates, and to do this efficiently, it constructs persons to whom communication is ascribed. In addition, since society needs a certain communicative contingency so that the system does not collapse into redundancy and static stability, it constructs persons as sources of communicative contingency, uncertainty, and restless construction of information. These persons, however, are not the psychic systems that Luhmann begins with when describing the situation of interpenetration. Individual psychic systems disappear the moment they open their mouths and say something, that is, try to introduce their internal complexity into each other. Interpenetration annihilates itself the moment it occurs. Interpenetrating systems are instantly transformed into the social system. This leads to the idea, which Luhmann explicitly suggests, that interpenetration describes not the *construction* of a system using the complexity of another system but the *emergence* of the social system from out of non-social psychic systems.

When we are talking about the social system, we are talking about the emergence of a system of a fundamentally different kind than anything preceding it. We have argued above that one of Luhmann's major theoretical innovations is that he understands meaning to be a level of emergent order beyond matter and life. If so, it would seem that interpenetration describes a situation in which a new type of system on a new and different level of emergent order appears. Luhmann explicitly compares the appearance of the social system from psychic systems to the emergence of meaning from life. The question, therefore, arises: Is the relation of the social system to psychic systems the same kind of relation as that of meaning to life? Does it make sense to compare the emergence of the social system from psychic systems, which, according to Luhmann, are both meaning systems, to the emergence of meaning from life? To say that life makes its complexity available for the construction of a meaning system is, of course, in a very abstract sense, plausible. Complexity, as we have seen in the discussion of the self-organizing universe, seems to give rise to emergent phenomena. Emergent phenomena are those that cannot be predicted and cannot be reduced to or deduced from what precedes them. When discussing the relation of psychic systems to social systems, we are not talking about the emergence of meaning from non-meaning.

Let us take a moment to reflect on emergence. The emergence of life from matter or meaning from life cannot be conceptualized as the communication of internal complexity from one system to another. Information is not being transferred from one system to another. Instead, an entirely new form of order arises (emerges) to reduce the complexity that the existing codes can no longer efficiently reduce. General systems theory allows us to think of matter and life as forms of systemic order. We also know that systems are dynamic; they operate and have agency, and therefore, talking about one kind of system "making its complexity available" for constructing another kind of system is theoretically consistent in an inclusive and general sense. Furthermore, if we accept the principle that every higher level of emergent order "integrates" the lower levels into itself, that is, life takes matter into its processes just as meaning takes life and matter into its operations, a certain mutuality or exchange of information is conceivable. The idea of integration, however, is not interpenetration because it is life, once emerged, that takes information from physical systems according to its specific coding, just as meaning, once emerged, takes information from life according to its semiotic coding. The emergence of semiotically coded meaning from genetically coded life cannot be understood as a situation of interpenetration. The relation of psychic systems to society is not a relation between systems that are environment for each other. Individuals are not integrated into society; they are constructed by society for the purpose of reducing the complexity of communication. The idea of interpenetration implies that the two systems mutually make their complexity available for constructing each other. Thus, it is a matter of mutual enabling, a co-evolution, whereby the two systems do not merge into a single comprehensive system. "The interpenetrating systems remain environment for each other" (214). They function as a source of "incomprehensible complexity" (214) for each other in a way that Luhmann explains in reference to Heinz von Foerster's "order from noise" principle.

> Thus, one could say that psychic systems supply social systems with adequate disorder and vice versa ... Social systems come into being on the basis of the noise that psychic systems create in their attempts to communicate. (Luhmann 1995:214)

Apart from the conceptual difficulties with the notion of any kind of relation between autopoietic, self-referential systems, which is not structural coupling, the recognition of the other as alter ego, which is the basis of double contingency, is sufficient to open up the social level. And it is difficult to imagine how or why one would recognize an alter ego if alter didn't communicate in some way, that is, if both ego and alter ego were not already within a social horizon of meaning. Double contingency explains why the interaction of ego and alter ego establishes mutual expectations, such as rules that social participants should follow normatively. In fact, when Luhmann explains how psychic and social systems can provide complexity for one another, it amounts to the establishment of norms, that is, "binary schematisms" (229) such as "conformity and deviance" (229) or right and wrong.

> A system may schematize the use of another system's complexity as friendly/unfriendly, true/false, conforming/deviant, useful/harmful, or whatever it wants. The schematism itself forces the system to admit the contingency of its behavior and thus the autonomy of the other system. … (230) For the interpenetration of human beings and social systems, this implies that the social meaning of an action is judged primarily by whether it corresponds to the norm or not. (233)

Friendly or hostile, right or wrong, conforming or deviating are not purely private states of consciousness but communicative acts. What is "provided" here as the complexity of consciousness for the communication system are precisely not states of consciousness but rather the actions of communicatively constructed "persons" in society. Schematisms or criteria thus do not have the function of admitting something merely conscious and non-communicative into the communication system, but instead, they subsume communicative complexity under certain expectations or criteria; they include and exclude; they constitute the ways in which contingent and unforeseeable communicative actions can be made meaningful. Schematisms are expectational and evaluative structures of the communication system whose function is to reduce communicative (rather than conscious) complexity and to permit connecting operations in the communication system. Somebody does something wrong and is blamed. Somebody does something right and is praised.

Luhmann admits that one of the consequences of interpenetration is that psychic systems now appear as *subjects*, "…binary schematisms are the precondition for the emergence of the figure that in modern philosophy has gone by the name of the subject" (233). The subject is characterized as "having true and false opinions" and "acting correctly and incorrectly or morally right and wrong" (233). These are the socially constructed persons we referred to above. It seems, then, that Luhmann is merely reintroducing double contingency here under the title of interpenetration. The application of binary schematisms runs through interactions and mutual observation and serves to specify and build up structures of expectation. If interpenetration is to answer the question of the possibility conditions of double contingency, then double contingency is used to explain itself. However, as already shown above, this model is only useful within the communication system – where commonly recognized criteria/expectation structures are already present – for the account of the concretization and *differentiation* of communication and not for the explanation of the emergence of a social system from pre-communicatively conceived systems of consciousness. Can there be any sense in judging actions as right or wrong if the agent – as a non-communicative consciousness – could not be told this, i.e., if they could not have their actions constituted by intersubjective control? If double contingency is to serve as a model of the specifically social, then the social must be the precondition and the condition of the possibility of double contingency. One can only get out of this circle if one admits from the outset that meaning emerges based on semiotic coding and that there is no fundamental difference between psychic and social systems. This does not preclude that a meaning system such as modern Western society chooses to differentiate itself into individuals and groups.

Finally, it would also have to be considered that the model of double contingency cannot be justified by the modal logical notion of contingency anyway. In the case of double contingency, it is not a matter of merely being able to be different but of disappointable structures of expectation. The difference between environment-1 and environment-2, i.e., between indeterminate and determinate complexity, is ignored if one interprets the phenomenological possibility horizon in a modal logical way. Modal logical contingency is not the contingency that the idea of double contingency presupposes. Modal logical contingency does not exclude the indeterminate arbitrariness of environment-1, whereas the model of double contingency presupposes communicatively applied symbolic generalizations, i.e., the already reduced contingency of environment-2. The children on the playground are not somewhere in chaos. They know quite well where they are and what they can do, which is why there can be a right and wrong way to do it. Ego expects a particular response to its communications from alter. This expectation may be disappointed. As a reaction to this, ego may correct his/her communications. As argued above, we are not dealing with coordinated behavior or structural coupling but with mutual understanding based on communication and language.

In our opinion, Luhmann's orientation towards Husserl's phenomenological analysis of meaning, on the one hand, and the Parsonian model of double contingency, on the other hand, blocks the development of an adequate theory of meaning based on systems theory. Contrary to Luhmann's view, the modal logical concept of contingency, by which he wants to link phenomenology with sociology and thus avoid the traditional problems of empirical versus transcendental cognition, can neither overcome the solipsism of transcendental phenomenology on the one hand, nor the biological reductionism of the double contingency model on the other. Only communication can reduce contingency to meaningful contingency. Thus, meaning exists only under the conditions of communication. There is only one system of meaning, society, consisting of persons and many other entities, depending on how the meaning system differentiates itself internally.

2.8 Communication

Within the framework of a general systems theory, we may thus postulate that the elements of a system of meaning are not adequately grasped merely as operations of purely psychically understood observation or self-observation but only as semiotically coded "communication." This theoretical decision requires that the concept of communication carry an enormous burden. It demands a theory of communication that goes far beyond what the tradition has previously offered. Systems of meaning are systems of communication. But what is communication?

Following the various theories of communication (Shannon/Weaver 1949; Wiener 1948; Bateson 1972 and others), the concept of communication is often taken to be so broad that it can denote any interaction at all. For example, Sebeok (1991:22-3), who advocates a "pansemiotism":

> All living things – whole organisms as well as their parts – are interlinked in a highly ordered fashion. Such order, or organization, is maintained by communication. Therefore, communication is that criterial attribute of life which retards the disorganizing effects of the Second Law of Thermodynamics, that is, communication tends to decrease entropy locally. In the broadest way, communication can be regarded as the transmission of any influence from one part of a living system to another part, thus producing change. ... An implication of this way of looking at communication is that the capacity for message generation and message consumption, which are commonly attributed only to humans, is here assumed to be present in the humblest forms of existence, whether bacteria, plants, animals, or fungi, and, moreover, in their component parts, such as subcellular units (for example, mitochondria), cells, organelles, organs, and so forth. The global genetic code, too, can (as it has been) quite fruitfully be analyzed in communicational terms: the message originates in a molecule, the master blueprint called DNA, its end being marked by a protein. The intricate interplay of nucleic acid and protein, the essence of life on earth, provides a prototypical model for all forms of communication.

Because of the danger of reductionism as well as the blurring of the essential difference between signal and sign associated with this view of communication, as well as the problems with covering over discontinuities in levels of emergent order Köck's (1987:359) critical reservations are to be agreed with:

> It is quite obviously unproductive to call any interaction 'communication' that may occur anywhere in the universe or in or between inanimate and/or animate bodies. ... It seems to

> me, therefore, reasonable and useful to understand the term 'communication' as a designation of a class of specific intentional interactions between living beings, namely those interactions which take place mediated, via media, i.e., with the help of signs, more precisely: whose necessary condition is the existence of a code, by which a (possibly, as in 'natural' languages, structured in itself) set of signs is linked with the corresponding set of meanings, without saying anything about sign modalities, stability of the code and the like. One could thus also speak of 'semiotic' interactions of all kinds.

Let it be emphasized again that it is not our job to deny the fruitfulness and scientific validity of research approaches that consider inorganic and organic processes as "communication." We do not in any way deny the legitimacy of "physiosemiotic," "phytosemiotic," "biosemiotics," and "zoosemiotic" approaches. It should be clear, however, that the concept of sign thereby loses any unified meaning and disintegrates into mere equivocation. Thus, to avoid the danger of a biologistic reductionism in social systems theory, we recommend, in agreement with Eco (1972:31ff), to postulate a "semiotic threshold." According to Eco, phytosemiotic and zoosemiotic research are

> ...to be identified as a kind of lower limit of semiotics, as the point at which semiotics emerges from something that is not semiotics, as the connecting link between the world of signal and the world of meaning – as in physical anthropology between the last primate and homo sapiens.

Plants, insects, and animals that coordinate their behavior with each other via the exchange of genetically encoded signals are not elements of a meaning system, that is, as long as the sciences that study these phenomena are not the object of study. In terms of systems theory, communication, in the restricted sense we propose here, functions as a reduction of complexity by means of a boundary-drawing system/environment difference that constitutes an autopoietic, self-referential, operationally and informationally closed system at the specific level of emergent order called meaning. Thus, a semiotically organized meaning system does not consist of plants or animals that coordinate their behavior with each other by means of structural coupling – although, of course, these are described by a meaning system; a meaning system does not even consist of "rational animals," even it prefers to differentiate itself by this term, but it consists of meaning and information. Man, as Peirce said, "is" a sign.

A second important consequence of the view that meaning systems consist of meaning or information and that meaning and information must be thought of as communication is that we must think of a systems theory in the social sciences as a communication theory. The leading question of a social systems theory is: How is communication to be analyzed in terms of systems theory? It will become apparent in what follows that a systems-theoretically oriented theory of communication has to deal primarily with three problems: 1) the problem of the subject of communication, 2) the problem of communication and action, and 3) the problem of the unity and differentiation of communication. We turn first to the problem of the subject of communication, that is, who is it that "speaks," and related to this, we will discuss the problem of the identity and/or difference of communication and action.

2.9 Communication, Subject, and Action

According to Luhmann, communication is the distinguishing feature of social systems. Therefore, he must offer a theory of communication that does not conceptualize communication as the action of individuals or understand communication as one possible social activity among others. If communication defines the social, it cannot be something that people could do but choose not to do as they wish. If the meaning system is the social system and the social system is a system of communication, it cannot be the case that individuals can decide to communicate or not or do something else such as hunting, gardening, building a house, etc. These are actions that Habermas, for example, would call "instrumental" as opposed to "communicative." Not only does the action theory model of society undermine systems theory, but it also undermines the primacy of meaning for the social. It undermines systems theory because a society that consists of individual subjects is not a system but a group, a collective whose order must be somehow derived from the actions of the individuals and not by processes of selection, relationing, and steering at the systemic level. This is what Hobbes did by explaining how the Leviathan was created by the social contract which integrated individuals into the state. From the beginning of the discipline, how to integrate individuals into society has been the fundamental question of sociology. Luhmann emphasizes that the theory of social systems does not ask this question since society does not consist of individuals. From the point of view of systems theory, a society composed of individuals and their actions does not consist of its own dynamic operations. Therefore, it cannot be conceived of as a system. By insisting that society is a phenomenon that can best be explained as an autopoietic, operationally and informationally closed system on the level of emergent order of meaning, Luhmann must propose a theory of communication that does not depend upon individual subjects and their freely chosen communicative actions. This is why Luhmann explicitly distances himself from the enthusiasm that classical cybernetics shared for Shannon and Weaver's transmission model of communication. The transmission model sees communication as a linear process initiated by a sender that sends information over a channel to a receiver. Instead, Luhmann describes communication as a self-referential, circular process, "a synthesis of three selections, as a unity of information, communication, and understanding" (1995:147).[114]

The transmission model of telecommunication and computer science also has the problem that it has nothing to do with semiotic coding. Shannon and Weaver posed the question of the most efficient technological transmission of electronic signals. According to this model, information is reified into a thing-like identity transmitted through a channel from sender to receiver. Luhmann rejects this view of informa-

[114] With this model Luhmann also wishes to distinguish systems theory from subjectivist positions that understand communication as "action, speech, proclamation, utterance" (147) or as some kind of action initiated by a subject.

tion for systems of meaning since it makes communication a process completed when the information is received. For systems of meaning, on the contrary, communicated information is necessarily interpreted and interpretable. It must somehow come back to the sender so that one communication can connect to another and so on, thus guaranteeing the autopoiesis of the system. The open world horizon of references Husserl describes serves as the context of any selection of information. Whereas signals trigger behavior, signs must be "understood" within a world of meaning. It is remarkable that although Shannon and Weaver themselves clearly denied that their model applied to semantic meaning, almost all communication or information theory in all areas of physics, biology, and sociology is based on this machine model. Meaningful information, however, has no substantial identity that can be sent through a channel to a receiver with minimum distortion but is continuously transformed in the communication process, a process that is completed, as Luhmann argues, only when it refers back to itself in a circular way such that ego knows that alter ego has "understood" and responded which makes further communication possible. Therefore, communication includes a third selection beyond merely utterance and information, namely, the selection of understanding. The selection of understanding, that is, the circular self-reference of communication, closes the loop and leads automatically from one communication to other communications and thus to the ongoing autopoiesis of the communication system. This means that semiotic coding requires a specific selection and relationing process that cannot be modeled according to the encoding/decoding model of machine signal transmission.[115]

Looking now more closely at the transmission model, if someone sends something to someone, e.g., a man sends a lady flowers, the only relevant question for the transmission model is how many flowers, how quickly, are transmitted without damage. This is it. For social communication, on the contrary, the relevant question is how the lady receives

---

115 Nonetheless, one idea that Shannon and Weaver introduced into communication theory has significance beyond mathematical information theory, namely, the idea that information has something to do with improbability and negentropy. We will return to this idea below when discussion information and networks.

the flowers and what she interprets them to mean. The information that the flowers are intended to communicate, and the expected response are not the flowers themselves but what they are taken to mean and how they are interpreted. The man no longer has the flowers but still has the information. Luhmann rejects the entire "thing metaphoric" (139) of the transmission model, which suggests that information is the same for the sender and receiver. The lady may "understand" the flowers in many different ways and respond unexpectedly. The recipient selects an interpretation from among several possibilities. According to Luhmann, this is where communication actually happens. And there is no communication if it doesn't happen. If the flowers are lost on the way, no communication has taken place.

Luhmann's model reverses the order of sender to receiver and understands communication as a form of circular causality.[116] Only when the response comes back from the receiver to the sender has communication taken place. The sender selects, for example, certain flowers as information which at a certain time are then "sent" in various possible ways – let's say red roses instead of white lilies, on their wedding anniversary instead of to a new acquaintance, and through the post instead of bringing them himself. But there is no communication until the lady somehow responds and says "thank you" or sends the flowers back or ignores the communication, which is also a response. As Watzlawick pointed out, one cannot not communicate. Social communication is a "selective process" (140) and not a technical transmission problem. The process includes three different but related selections. First, a piece of *information* is selected, for example, the flowers; second, a form of *utterance* – Luhmann also speaks of a "*Mitteilungsverhalten*" (196) – such as sending the flowers by post or currier, or attaching a note, or calling, etc., and third, the communication must be recognized as such. The lady must receive the flowers and *understand* this as a communication and thus respond in some way. Recognizing a communication means understanding it as a communication through an "acceptance selection," which could also be rejection that manifests itself in a subse-

[116] "Communication is made possible so to speak, from behind contrary to the temporal course of the process" (143).

quent response. All these three selections of information, utterance, and acceptance/understanding must happen before communication in the complete and relevant sense of the word can take place.[117] Communication "only comes about when ego establishes its own state on the basis of a communicated information," which implies that alter ego is actually the sender, or at least that it is a circular relation. Double contingency implies this circularity, which in turn results from ego recognizing alter as alter ego and vice versa. As Luhmann argues, this is based on distinguishing between utterance (the act of communicating) and information (that which is being communicated).

If all three selections are not present, the attempted communication has failed. The addressee did not hear or heard differently, was unaware that anyone was trying to communicate, did not consider the speaker capable of communication, etc. Communication can only occur in a situation of double contingency, that is, mutual social recognition, which is a circular, self-referential situation. If there is no expected behavior on both sides, such that the sender is also the receiver and vice versa, there are no possibilities of connection to further communication. Except, of course, to try again. It is important to note that "understanding" is not described as an intentional act of a subject but as a moment in the systemic processing of communication. If the flowers do not result in a reaction, if one does not even get a thank you note, then the communication has either not been established or the interaction continues due to other connecting or associated messages. One calls and asks, for example, whether the flowers have arrived. Understanding as the third selection thus has the function to enable connecting or associated further communication possibilities and so on such that the autopoiesis of the social system is maintained. For the autopoiesis of the communication system to be maintained, communications must connect to

---

117 Luhmann (1995:147) adds a fourth selection which he calls "acceptance or rejection" which is conceptually hardly distinguishable from understanding but serves the purpose of enabling the development of certain forms of communication which Luhmann, following Parsons calls "symbolically generalized media" that influence participants in favor of acceptance. These media become the basis for the differentiation of the semi-autonomous functional subsystems which modern society, according to Luhmann, consists of.

further communications. Someone says something, and someone else responds in some way, and so on in one unbroken chain of communication that makes up the social system. The meaning system, therefore, is not composed of thoughts, feelings, or actions but of a chain of communications. One communication is networked with the following one by the connecting behavior, which alone provides information about whether one has been understood or not and whether one has understood oneself correctly or not. I can no more "look inside" myself than I can look inside another person to determine whether the words have been understood. In communication, communications do not operationally refer to people, inner mental states, or things or events in the world but self-referentially to communications.

According to Luhmann, communication is thus the indissoluble synthesis of the three selections: the selection of information, utterance, and understanding. Insofar as the unity of all three selections is only analytically resolvable, there can be no information in the operations of the communication system that is not communicated or could not be communicated; there can be no utterance that is not understood or could not be understood. And there can be no understanding of anything other than communicated information. Every selection of understanding is also automatically and necessarily a further communication. The ground of possibility of this synthesis of the three selections lies in "coding" (142), i.e., the specific organization of communication into a semiotically coded meaning system.[118] The smallest possible unit of communication is when someone says something to someone. This is a semiotically coded synthesis of the three selections and thus is constituted as a supra-individual interaction context that cannot be thought of as the achievement of consciousness, a subject, or an individual, or even as some collective action of individuals who choose to communicate with each other or not. Communication – and thus meaning in general! – is therefore never only mine or yours, but always ours, always social. If communication is the basic unit of society, and the selection of a subject is not part of

118 "The combination of information, utterance, and expectation of success in one act of attention presupposes 'coding'" (Luhmann 1995:142).

the threefold synthesis, then the subject or the human being is not an element of the social system, and the actions of individuals cannot serve as the basis for creating society. When Hobbes' free individuals in the state of nature agreed to sign the social contract, they must have been able to communicate and, through communication, understand themselves as "free individuals" – which is not a natural state but a social role! – who are authorized to enter into a social contract. Having thus gotten rid of the subject, the question arises: who communicates, who is it that speaks?

Interestingly, even though the communication synthesis as a whole is the basic unit of the system of meaning, which cannot be resolved into the actions of subjects, and which therefore does not imply subjects or individual speakers, at least not human beings, it still remains impossible for the communication system to dispense with the subject as the originator of communication. This is evident in Luhmann's sharp distinction between actions and communications on the one hand and his insistence that actions, not communications, constitute the elements of the social system after all. On the one hand, it is asserted (164) "that communication cannot be conceived as action, nor can the process of communication be conceived as a chain of actions," while on the other hand, we hear (165) that "communication systems...must interpret utterance as action" and that "a social system is constituted as an action system." If the word "action" here does not mean simply "operation" of the system but something that someone does, how can we understand this apparent contradiction?

According to Luhmann, the communication system itself must conceive communication as non-communication, i.e., as action, because the three selections of information, utterance, and understanding cannot be observed directly. Communication can be "only inferred" (164) through visible communicative behavior. We perceive not syntheses of three selections but people talking to each other, exchanging things, cooperating in various endeavors, and so on. In addition, there is the problem that communication is a "symmetrical relation" of the three selections, which leads to uncertainties and potential blockages that must be "asymmetrized" (165). "Communication is symmetrical insofar as each selection can lead the others and the leadership relations can be continuously reversed." Making an utterance, for example, can lead to looking for some information that one wants to say, and this, in turn, can lead to eliciting a response from someone, which can lead to saying something because someone said something to which one wishes to respond. The symmetry problem here is that one doesn't know which selection comes first, where to begin, and how to proceed. Is this situation not like the famous chicken or egg problem? Where should we start? To solve this problem, it becomes necessary for society to *ascribe* communication to a speaker. Every communication also has an ascription to someone doing the talking.[119]

How should we order the communication process and keep it going if we don't know who said something first and who must respond or react? The symmetrical arrangement of selections creates an unacceptable complexity because everyone could be talking and responding about everything in different ways simultaneously. In the face of this problem, the communicative process threatens to end. This problematic complexity can only be reduced by directing communication along certain lines. This is done through the attribution of actions:

> Only by building the understanding of action into the communicative occurrence does communication become asymmetrical, only thus can a person who utters information give directives to its receiver, and this can be reversed only if the receiver begins to utter something of his own, that is, begins to act. (Luhmann 1996:165)

Communication, therefore, seems necessarily to become communica-

[119] The theory does not specify who the speaker (communicator) is. This depends of the available forms of self-reference within the system. In principle anything can communicate, not only human beings, but animals, ETs, spirits, God, indeed anything. This marks the post-human position of Luhmann's theory. Human beings are not only banned from the social system, but they are also not necessarily reconstructed within it as the only initiators of communication. The theory requires an ascription but says nothing about what it is to which communication is to be ascribed.

tive action. According to Luhmann, however, communication in itself has nothing to do with action because communication is a synthesis of three selections, and action is something that can only be related to one selection, the selection of utterance. Luhmann's theory implies that information has no agency of its own. Contrary to Bateson's definition of information as a difference that *makes* a difference, for Luhmann, agency lies solely on the level of the system as a whole; only the communication system does something, and it attributes its processes to a speaker. Acting human individuals are banned from the system. Still, they are reintroduced into the communicative processing of meaning by interpreting communication selections as communicative actions and thus constituting them as being able to be connected to other communicative actions in an ongoing chain. Communications are attributed to actors or, as Luhmann puts it, "persons" who are communicatively constructed by the social system precisely for this purpose. Thus, it is communication that constructs persons and not vice versa. Actions of persons are the "attributions" of communications necessary for the connectivity of processing meaning. Without the system being able to attribute communications to "someone," the system would remain unable to conceive of communication as an operation of an operator at all and thus to refer to itself on any other level than the metaphysical level in which every communication would only be attributable to the system as a whole. It is on this level that Luhmann can say that communication communicates.

In general, all autopoietic systems are necessarily self-referential. Without basal self-reference, the system cannot refer its operations to its own operations. Self-reference is a necessary prerequisite of all autopoiesis. As noted above, the specific self-reference of a system of meaning occurs through reflection on system identity, i.e., the system maps its system/environment difference onto itself. This creates what we have called a metaphysical or world-inclusive self-reference. The autopoiesis and operational closure of the system require that the system refers to itself as doing something, as a dynamic entity. The world is, therefore, always a "product," even if we prefer not to speak of a creator God but of a self-organizing evolutionary process. If we do not want to attribute all agency to God, it becomes useful to differentiate further self-references within the system, that is, to identify speakers. Without being able to answer the question "Who communicates?" on the level of everyday life, processing communications and linking one communication to another would be challenging. It would be as though God would have to do everything; even the most minor decisions and events would have to be authorized and carried out by God. Some believers try to live in this way; they leave everything to God, but these people usually can only survive in very small communities of the like-minded, that is, of those willing to wait until God has cleared everything so that life can go on.

Internal differentiation of self-reference requires that the system becomes the subject for itself on levels below the totality. The system must make "a description of itself ... in order to control the progress of processes, the reproduction of the system" (165). It does not matter who is speaking so long as someone is, and an ascription can be made. There could be humans, gods, spirits, ancestors, animals, plants, nature, the Freudian subconscious, and beings from other planets, but also the state, reason, conscience, law, justice, fate, the stars, invisible voices, etc. who all can appear as "persons;" the only important thing is that a system reference exists and that an attribution of action is possible.

The communication system is indifferent to holders of the subject position; all that matters is that the position is filled and there is a – any – subject. The requirements of self-referentiality and operational closure of the communication system thus create the appearance – Luhmann speaks of a "self-simplification" – of a system of action and constructs the actors necessary for it, whoever they are: "For purposes of self-observation and self-description, the symmetry of communication is asymmetrized..." (165). Acting subjects are thus constructed as elements of the social system and simultaneously distinguished from communication since they can choose to communicate or not. As well shall see, society must take account of this by constructing many supporting structures to ensure the motivation and cooperation of such actors; among the most important of these are "symbolically generalized media," which lead to the self-organization of autonomous functional subsystems, which we will discuss below.

In summary, Luhmann's communication theory is based on the idea of autopoietic, self-referential, and operationally closed systems. The system in question is the social system that appears on the level of emergent order of meaning where it is meaningful information that is communicated in circular, self-referential operations that are a synthesis of the three selections of information, utterance, and understanding, and which is confronted with the problem of reducing complexity by means of internal differentiation into actors or subjects of various kinds. Much of this theory is quite useful for understanding communication as a systemic form of order. The question arises, however, of whether Luhmann is not unnecessarily subscribing to a minimal notion of agency. Because of the unnecessary and, in our opinion, even untenable distinction between psychic and social systems, Luhmann puts himself, it would seem, into the situation of excluding individuals and their actions from society while at the same time having to reintroduce them into society through the back door as socially constructed persons who alone can initiate communication. What is the relation, we may ask, between psychic systems and those actors who, as persons, populate the social system? How can the one be transformed into the other? For Hobbes, this was a simple matter of signing the social contract. For Luhmann, however, the system/environment difference must be overcome since psychic systems are banned from the social system and find themselves in its environment. Since there is, in principle, no communication between system and environment, how are psychic systems transformed into elements of the social system?

The first answer to this question is to simply refer to the general principle of systemic order that all systems construct their own elements. According to its organizing principles, the table constructs a top and legs out of whatever material is at hand or desired. An organism constructs tissue, organs, etc., according to its genetic coding from the material substrate. A meaning system also constructs its elements. These are information as processes of communication. This is called self-organization. The environment offers systems the problem of complexity that they must solve. Luhmann follows Heinz von Foerster's notion of "order from noise" to understand this process. Generally, when systems are environment for each other, they are nothing other than environment and can only perturb each other, that is, offer contingency, indeterminacy, variety, and complexity so that the system can exorcise its self-organizing power and create, as von Foerster put it, order from noise. Luhmann, as we saw above, calls this "interpenetration." Colloquially, one could say that individuals make life complicated and unpredictable for society, and society constantly makes life hard for individuals. This traditional sociological commonplace is reinterpreted within systems theory as interpenetration. It is the systems theoretical version of the modern Western paradox of free individuals vs. constraining society, the one vs. the many, the individual vs. the group, or the classic puzzle of how to integrate individuals into society while at the same time allowing sufficient freedom to prevent society from collapsing into sterile conformism.

In the systems theory version of sociology, the social system uses the contingency or "noise" generated by psychic systems to construct order. On the other hand, psychic systems use the constraints of society to construct deviation, contingency, freedom, and indeterminacy, which is the unique way in which they self-organize. The general principle at work here is that systems self-organize in response to complexity. Psychic and social systems "interpenetrate" in so far as they mutually introduce contingency into each other and thus instigate self-organization. Why the social system needs the contingency of psychic systems instead of the contingency of the world, or in other words, the question of whether the world itself is not sufficiently complex and thus able to introduce all the indeterminacy any social system could ever want completely independent of individuals is not addressed by Luhmann. One could suppose that the sociological interpretation of the system of meaning as society and not as world leads Luhmann into steering the systems theory of meaning into a sociological corner where questions of the self-organization of meaning become automatically questions of how individuals relate to society. Perhaps an adequate theory of systemic order on the level of meaning must fundamentally revise the assumptions of sociology and look for other foundational concepts than "society."[120]

The above discussion allows us to ask why agency in Luhmann's theory must be limited only to the utterance selection and have nothing to do with the selection of information. Ascribing agency only to utterance seems to be a consequence of the aim to banish subjects from society, and indeed from the theory of society as a system of communications as well. But apart from the desubjectification program Luhmann embarks upon and the somewhat questionable reintroduction of subjects as speakers, one can still ask why information is devoid of agency. If information is defined as differences that "make" differences, then where does this efficacy and agency to make differences come from? How does information "make" a difference without any agency of its own? And what kind of agency could information have? Is it impossible to suppose that meaning is a dynamic process that arises out of information? Could it not be that it is information that "selects" agents and speakers? Did Luhmann himself not say that the subject is not primary since meaning is the condition of its possibility? As argued below, this is an option that the network paradigm, in contrast to the systems paradigm, explores. We will ask whether systems theory can also accommodate the agency of information or whether it is precisely why systems theory tends to break down and must be extended by network models. But we must not get ahead of our story.

As a response to the question of agency and information, one could argue, as Luhmann does, that differences can only make differences insofar as they are communicated and understood. And, as we saw, communication requires actors who communicate. But if one says the same thing over and over again, there is no information, and no difference is made. Redundance has no informational value, regardless of how much it may be communicated. Utterance does not make the difference that information can make, for there can obviously be a lot of

---

[120] This is precisely what we will suggest when discussing the network paradigm below.

communication – we experience this daily – without much information. But when information is communicated, it makes a difference because it has its own power to change reality quite apart from the utterance. If one understands, for example, that tobacco, alcohol, butter, and frozen meat are dangerous to one's health, then, as Luhmann (203) notes, one has become "another." It can no longer be ignored. In what does this understanding show itself, if not in some connecting behavior that gives others and me an understanding of my understanding? How is understanding itself to be understood if not as a state-transforming operation that can somehow be attributed to information and not to a subject or individual actor? Philosophical hermeneutics does just this by attributing understanding to the efficacy of a text, an artifact, or a historical event. The efficacy of information goes beyond the author and interpreter. It arises from the text or artifact that itself has the power to open up a horizon of meaning. After all, where do the many "references" come from that Husserl already identified as meaning? Philosophical hermeneutics emphasized against Husserl that understanding is a historical process and not an act of an autonomous subject. Gadamer (1960), for example, takes his cue from theological and legal hermeneutics, where understanding the Word of God or the law manifests itself directly in practically changing the situation of the interpreters and thus realizing the historical process.[121]

Given these considerations, it would seem that agency could be conceptualized as distributed throughout the system, attributable to all system elements and operations. Luhmann chooses not to pursue this path when conceptualizing the system of meaning. Instead, a speaker is constructed by communication for attribution. These speakers are conceived as socially constructed "persons" who replace the human individuals of traditional sociology. Luhmann is forced to relate action to only one of the three constitutive selections of communication, and this is only a kind of necessary evil to make communications asymmetrical and ordered in time. This is a requirement of the temporalized processing of communication and has nothing to do with traditional

[121] For a discussion of philosophical hermeneutics in relation to systems theory and network theory see Krieger/Bellger (2014).

sociological theories of agency vs. structure.[122] Nonetheless, it seems to be a truncated view of agency since, in effect, it introduces human individuals into the social system through the back door. Theoretically, this is not necessary. It is possible to argue that agency is implied as much by the selection of information and the selection of understanding as by utterance and that the subject, therefore, does not need to be reintroduced into systems theory as a privileged source of agency.[123] Despite these objections to Luhmann's theory, we readily admit that replacing the act of speaking with the systems theoretical notion of selection and operation is a significant theoretical innovation in sociology that must be kept in view. The system is the actor![124]

2.10 The Differentiation of Communication: Functional Subsystems

For Luhmann, a systems theory of meaning is a theory of the social system as a system of communication. As a system of communication, the social system can only become internally differentiated by means of differentiating communication. Luhmann succinctly puts it: "The differentiation of social systems can emerge only through the differentiation of communication processes" (1995:152). It is not semantic content of any kind that is decisive for meaning; it is how communication of any content is structured. It is not what is said but how it is said that differentiates society. How does the social system differentiate communication? As the answer to this question, Luhmann proposes his definition of modernity. Modern society differentiates communication into *functional subsystems,* from which he concludes that modern society is a functionally differentiated society:

> Functional differentiation means ... that autonomous subsystems of society are formed in the orientation to their own respec-

[122] See Giddens (1984) for an influential statement of this position.

[123] Contemporary non-Cartesian cognitive science (see Rowlands 2010) and theories of distributed agency have persuasively shown under the name of "affordances" that things also have agency and contribute to cognition.

[124] We will return to this slogan in the discussion of actor-networks below.

> tive functions, which reproduce themselves self-referentially, orient themselves recursively to the respective self-produced communications, and thus realize the characteristics of structurally determined autopoietic systems. (Luhmann 1990b:479)

The social subsystems Luhmann has in mind are the economic system, the legal system, the political system, education, art, religion, science, the media system, and healthcare, to name only the most important ones. These subsystems serve specific social functions and can "reproduce themselves self-referentially" and "orient themselves recursively to the respective self-produced communications." They can do this because they operate autopoietically, self-referentially, and are operationally and informationally closed based on their unique *binary coding*. Such binary-coded functional subsystems process all communication in society. After all, if one is not engaged in business, education, and training, political activities of some sort, healthcare, art, religion, science, media, etc., what is one doing? Perhaps there is room for small talk, but this is also systemically organized as "interaction systems." As sociology (see Gofman) has long known, socially constructed persons in the modern functionally differentiated society are like actors playing roles in different "frames," settings, or contexts according to varying scripts for various purposes. According to Luhmann, the codes that structure this communication in one way or another emerge in society from the differentiation of *symbolically generalized media*.

Media, in general, are "evolutionary achievements that start at ... ... breaking points of communication" and serve to "transform the improbable into the probable" (Luhmann 1984:220), whereby the improbable here consists first of all in the fact that communication occurs at all. For Luhmann, each of the three selections constitutive of communication mentioned above (information, utterance, understanding) is improbable. That one has something to say (information), that the message reaches someone (utterance), and that one is understood and responded to are all improbable. Assuming these three improbabilities, three media function to make the improbable probable: Language, dissemination media (such as the press and radio), and finally, what

Luhmann, following Parsons, calls "symbolically generalized media." On the concept of symbolically generalized media Luhmann writes (1995:161):

> We would like to call "symbolically generalized" the media that use generalizations to symbolize the nexus between selection and motivation, that is, represent it as a unity. Important examples are: truth, love, property/money, power/law; and also in rudimentary form religious belief, art, and, today, standardized "basic values."

Symbolically generalized media, as opposed to other media such as language and dissemination media such as writing, radio, the press, etc., play a major role in differentiating social subsystems.

> In all these cases ... it is a matter of conditioning the selection of communication so that it also works as a means of motivation, that is, so that it can adequately secure acceptance of the proposed selection. The most successful and most relevant communication in contemporary society is played out through these media of communication, and accordingly, the chances for the forming of social systems are directed toward the corresponding functions. (161)

Luhmann ultimately attributes the internal differentiation or build-up of internal complexity in the communication system via such symbolically generalized media to the contingency of the environment, namely to the assumption, already discussed above as problematic, of a multiplicity of pre-communicative consciousness systems that think and act "obstinately" (1990b:40), and which would tend not to participate in communication at all if special motivational structures (namely: the symbolically generalized media) did not facilitate and steer their integration into the social system. What for Hobbes was accomplished by the need to secure one's life in the war of all against all is, for Luhmann, accomplished by symbolically generalized media that "motivate" people to communicate and cooperate.

According to Luhmann's functional differentiation model, interactions or communications are absorbed into a particular symbolically generalized medium where a functional subsystem of society codes them to make it much easier to "understand" what the communication is about. The codes of the functional subsystems are two-valued or binary codes.

> The most important functional systems structure their communication by means of a binary, two-valued code that claims universal validity from the point of view of the specific function in each case and excludes third possibilities. (Luhmann 1990:75)

Money, for example, is seen as a symbolically generalized medium in which a certain communication is made possible by attributing a monetary value to something. After something has its "price," all that remains relevant communication is whether to buy or sell. The money code or the binary alternative of buying or selling filters the world into only two values: buy or sell. All other communications, such as those about religious belief, educational certification, legal judgment, political office, artistic value, etc., are excluded from the economic system. The binary coding of a functional subsystem is both inclusive and exclusive but, paradoxically, for this very reason, universal. There is nothing that doesn't have its price; everything comes under the binary coding of the economic system. But only with regard to buying and selling. Everything else is banned into the environment of the economic system and is not registered by the code. One may discuss the aesthetic value of a work of art as long as one wishes, but at a gallery, one has to decide whether to buy the painting. Since a price can be given to anything, including love, the economic system becomes an autonomous, autopoietic, operationally and informationally closed system.

The same happens for the other functional subsystems. Due to the medium's claim to universality and the binary code, in which all interactions are structured by either the positive or the negative value, connecting up communications to one another becomes much more straightforward than in a situation where one must first clarify what is being said. Knowing, for example, that a particular communication is an offer to sell something and not the profession of a belief, an educational instruction, a legal argument, etc., makes it much easier to "understand" what is going on and communicate reciprocally. All communications relating to buying and selling are connected, and all other communications are excluded. The code enables the construction of a system/environment difference and, thus, the emergence of the economic system as a social subsystem organized by its unique code. Luhmann calls the sum of communications coded according to the binary scheme of money, that is, paying or not paying, the economic system. Due to the all-inclusive binary code, the system differentiates itself from the overall social communication as an autonomous, autopoietic, operationally and informationally closed subsystem, whose function is to facilitate the material reproduction of society, that is, the production and distribution of goods and services.

Corresponding to this model of the economic system, other functionally differentiated subsystems emerge within society out of their own symbolically generalized media. The political system is coded by the binary alternatives of either power or no power: disposition over offices and positions in the medium of power. The legal system is coded in terms of legal/illegal in the medium of legal decisions. The science system is coded in the binary alternatives of true/false in the medium of knowledge. The educational system is coded in the alternatives of certification or non-certification (pass or fail) in the medium of career expectations. And the religious system, according to Luhmann, is coded in immanence and transcendence in the medium of faith. One could add the art system, which is coded by what is considered to be original and dysfunctional as opposed to what is functional and already known in the medium of contingency. Finally, the healthcare system is coded by the alternatives of being sick or healthy.

The functional differentiation of social subsystems occurs as a reduction of increasing social complexity. Society has developed in complexity from segmented (archaic) to stratified (high cultural) to modern functionally differentiated societies. Functional differentiation results "autocatalytically" (1995:190) from the autopoietic reproduction of the meaning system. The general "mechanism" of evolutionary differentiation specified in relation to communication is called, according to Luhmann (1990b:49,81), "surplus generation-and-selection." What at the level of genetic organization provides for surplus and variety of living beings as random mutations is attributed at the level of semiotic organization to the arbitrariness of observation as it happens in the conscious environment of the communication system. "Distinctions are always arbitrarily introduced and used" (99). The uncontrolled observations of psychic systems introduce distinctions arbitrarily, creating a surplus of possibilities. Based on this variation, selections are made first by the symbolically generalized media and then increasingly by the various social subsystems. It becomes generally expected that everything be given a price and that markets are established, that everything can be either legal or illegal and that courts are established, that specialized competencies can be learned and certified by schools, and that true knowledge can be discovered by scientific methods leading to research becoming institutionalized. Furthermore, it becomes institutionalized that religion has to do with faith, not politics, which is organized in political parties and programs, and that healthcare has to do with doctors, hospitals, and health insurance and not with religious rituals. What comes out of this differentiation process is modern Western society.

According to the concept of autopoiesis, communications reproduce themselves in such a way that they simultaneously reproduce their own reproducibility, i.e., produce their own conditions of possibility. Since everything depends on the chain of communications continuing without breaks and gaps, the conditions of the possibility of communication can be found in whatever successfully ensures that communication is linked up to further communications. The symbolically generalized media and the functional subsystems ensure that the autopoiesis of the communication system can be maintained. They reduce the complexity of communication and make it manageable. The idea of differentiation means that systems reduce complexity by building systems within themselves.

Confronted with the imperative that the autopoiesis of communication must continue, the way to ensure communication becomes the construction of specialized subsystems within the social system.

> system differentiation repeats the formation of the overall system within itself. The overall system is reconstructed as the internal difference between a subsystem and the subsystem's environment, and this reconstruction is different for each subsystem. .... The overall system is contained within itself many times over. (1995:191)

Importantly, the internal differentiation process is not purposefully directed by the system as a whole. The subsystems that emerge within the social system have no special relation or informational exchange with the encompassing social system, which is for them nothing other than environment.

> internal system formation occurs autocatalytically, that is, by self-selection. Internal system formation presupposes neither "activity" by the overall system nor a capacity for dealing with that system, not to mention any overall plan. Nor does the overall system subdivide or break down into subsystems. The overall system merely enables the self-selection of subsystems through its own order. (1995:190)

Every symbolically generalized medium and the functional subsystems that emerge from them tend to absorb all communications into themselves according to their unique codes. In this way, the complexity of social interactions becomes reduced to those communications coded by the subsystem. Each subsystem must determine for itself what is system and what environment. The consequence of functional differentiation is that after the differentiation of such autonomous, autopoietic subsystems, no overall social communication remains. The various subsystems compete to represent the system as a whole. The economic system, for example, encodes everything exclusively in the alternatives of payment or non-payment, just as the legal system perceives everything as legal or illegal. For science, there is nothing that cannot become the object of research. For the religious system, the world is divided into sacred and profane. Everything else that is neither legal nor illegal, neither sacred nor profane, which has no price and cannot be bought or sold, which does not lead to an educational certification of some kind, etc., is banned into the environment.

In what larger communication contexts, then, are the respective subsystems contained? What code mediates communication between the functionally differentiated subsystems? How can overall social communication still be possible if the social system is fragmented into autonomous, informationally and operationally closed subsystems? On the other hand, it seems necessary to have overall social communication because if the semiotic code were not internally repeated and specified in the subsystems, we would not be dealing with a subsystem but with a completely different society. Each autonomous subsystem would be a different communication system, and there would be no encompassing social system that somehow contains the subsystems within it. Subsystems are subsystems only because they are in some way integrated into society as a whole. After all, they fulfill functions *for* society. This raises the question of the unity of society against the multiplicity of subsystems. Society becomes the environment for the functional subsystems, and each subsystem is environment for each other. How is the unity of the system/environment difference to be theorized for the system of meaning when society itself fragments into various autonomous subsystems? This is the problem of "ecological communication."

2.11 Ecological Communication or Can Society be Modeled as a System?

Conceptualizing the system of meaning as society and then going on to understand society as a system of communication that is differentiated into autonomous subsystems, each of which is only environment for the others, raises the question of the unity of society as a whole. Wherein lies the unity of a functionally differentiated society? In pre-modern societies, the unity of the "world" was religiously represented. It makes little difference that there were and still are different religions, worldviews, ideologies, cultures, etc. Religious and cultural pluralism does not amount to systemic pluralism. The pluralistic world is still one world; different convictions and views about the world give society more to talk about. In the modern functionally differentiated society, however, the "world" is no longer available since each social subsystem sees the world differently and blocks informational exchange between system and environment. When business people are involved in politics, science, education, or religion, they are not doing business. Modern functionally differentiated society requires that socially constructed persons are theatrical actors playing different roles in different dramaturgical settings for different audiences, etc. Goffman's dramaturgical role theory of social interaction describes this situation in detail. And, just as there is no unity for society as a whole, there is no unity for the actor. For a long time, modernity dealt with this problem by substituting "reason" or "rationality" in the place of God to represent the unity of the system of meaning. Not only has postmodern critique and deconstruction eliminated this form of metaphysical self-reference and raised the question of the extent to which any self-representation of society as a whole is still possible, but functional differentiation has replaced society with specialized forms of communication that are autonomous, operationally and informationally closed systems in their own right.

Is society anything other than the mere sum of the differences of the various subsystems to each other? What kind of communication thematizes the unity of the differences between the subsystems and their environment? This kind of communication Luhmann calls "ecological communication." It is misleading to assume that ecological communication talks about environmental problems such as depletion of natural resources, pollution, climate change, etc. Of course, this is what everyone means by ecology, and everyone expects communication about ecology to deal with these important issues. This is precisely what happens, but it occurs within the particular horizon of meaning of the various functional subsystems. The "natural" environment is a topic of laws and regulations, of business models, of scientific research, or religious beliefs and ethical admonitions, educational programs, art, political campaign promises, etc. But this is the problem and not the solution.

We must emphasize that we are not talking about ecological communication as it is commonly understood. We are not concerned with the "natural environment." We are concerned with the social environment, for even nature exists only insofar as it plays a role in communication and is constructed by meaning within society. We are concerned with the *unity of the difference between system and environment* for a system of meaning. Specifically, we are concerned with the problem of the unity of society as a whole under the conditions of functional differentiation, that is, under the conditions of a society that has differentiated itself into autopoietic, operationally and informationally closed subsystems such that each subsystem is related to the others and society – whatever that might still be – as to an environment. This is why the concept of ecology is of central importance to a theory of social systems. There may be no such thing as *the* social system, but only social systems that are more or less structurally coupled.

The attempt to conceptualize the social system under the conditions of functional differentiation brings the idea of ecology onto center stage. As we have already had occasion to note, ecology poses a serious problem for systems theory. On the one hand, every system is constituted by the exclusion of the environment, and on the other hand, ecology focuses precisely on the unity of the system and environment. As we have argued above, It would seem impossible to speak of an *ecosystem,* for this system would have to exclude an environment, which would need to be included again and so on infinitely or at least until one reached the entire universe. Ecology, therefore, stands for the problem of conceptualizing the unity of the difference between system and environment. This problem becomes acute when history arrives at functional differentiation as the primary form of social order. Luhmann (1989:5) speaks even of a "fundamental paradox" in "the theoretical structure of the ecological question." The unity of the difference between system and environment cannot itself be a system. What cannot be a system cannot become a topic for systems theory. Could it be that the unity of the difference between system and environment itself does not fall within the scope of systems theory? Systems theory presupposes it but begins only from the difference, not the identity. Could it be that the theory has a blind spot and cannot see what it cannot see? And if this might be the case, does this imply that society as a whole cannot be conceptualized as a system? If so, then meaning cannot be adequately modeled as a system.

In principle, there can be no communication, i.e., informational exchange between an autopoietic, self-referential, operationally and informationally closed system and its environment. Information is constructed within the system and not imported from the environment. Regarding information, there is no "instruction" but only construction, which occurs only within the system. The environment can only perturb the system but not inform it. Were this not the case, the "system" would not be autonomous, autopoietic, and informationally and operationally closed. Instead, it would be an element of a greater system, subject to the information constructed by the greater system of which it is an element. The tabletop, whatever material it might be made of or form it might have, is an element of the table system that constructs it as a tabletop. The organizing principle or structure of the table does not represent environmental complexity that pressures some piece of wood or metal to self-organize into a tabletop. This is also true for living systems. Meaning systems, however, have a unique relation to the environment.

Meaning systems have the unique property of constructing their environment as meaningful within their own boundaries. For this reason, it could seem plausible that functional subsystems within society can be "instructed" by the information already existing in the environment, which, since it is itself meaningful, must also be in some sense of the term "information" that can be communicated. The question of the unity of the difference between system and environment, the ecological question, would therefore be answered alone by the fact that for meaning systems, there is no non-meaningful environment. On the level of general systems theory, however, this solution leads only to the further problem of how a system constitutive difference between functional subsystems and their environment is conceivable. And if there can be no constitutive difference between system and environment, it makes no sense to speak of functional subsystems as systems in their own right. They must either be conceived of as system elements – which only pushes the problem of a constitutive distinction back to the social system as a whole – or, if they are autonomous systems, how can their unity with society be conceptualized? We will argue below that ecological communication is possible but not based on systems theory. Instead, we will say that meaning can only be adequately theorized as a network, not a system. In contrast, the functional subsystems within the social network can be modeled according to systems theory.[125]

Given these open questions, we must ask what the concept of the "social" means from a general systems theory perspective.[126] Despite all of Luhmann's claims to the contrary and the entire program of a theory of social systems, society as a whole seems unable to be modeled as a system. Various reasons support this claim. These are:

1. A distinction between system and environment does not constitute meaning and society.

2. Meaning is not subject to evolution. Even if one conceptualizes meaning as a system, the system cannot be understood as an autopoietic, self-referential, operationally and informationally closed

[125] The theory is not saved by simply thinking of meaning as a medium, as Luhmann suggests, since a medium is only environment for a system and not a system itself.

[126] The anomaly, that no communication can address society as a whole was recognized already in the 90s; see Fuchs (1992). Instead of being allowed to challenge the systems theory paradigm, however, the problem was dealt with in Ptolemaic fashion by generated ever more theoretical epicycles. Luhmann (2012:40) attempts to address this issue by distinguishing between three levels of analysis: "(1) general systems theory and, within it, the general theory of autopoietic systems; (2) the theory of social systems; (3) the theory of the societal system as a special instance of social system." This merely pushes the question back to the social system since society as a whole is a "special instance" of the social system. Luhmann attempts to resolve these issues by describing the operational closure of communication. The social, regardless of level of generality, is nothing other than communication. The concept of communication, however, depends on a theory of meaning which leads to a tension in Luhmann's theory between meaning and society.

system. This means that the relationship between the system and the environment cannot be understood as a reduction of complexity under the pressures of evolutionary selection.

3. Although social systems can emerge within meaning, there can be no social system, that is, no system of the whole of society, world, or meaning.

Let us look more closely at these three reasons for doubting the systems model of society.

1. Meaning is not Constituted by a Distinction Between System and Environment.

If meaning is to be modeled as a system, it must be constituted by a difference between system and environment. Processes of selection, relationing, and steering must reduce the complexity of the environment so that order is established against a relatively disordered, random, unstructured, or, at any rate, more complex environment. Of course, no system arises in a completely unstructured environment in the real world. If there were no regularity in the universe, no structure, no natural laws governing matter, energy, etc., then no kind of order could establish and maintain itself. For this reason, Maturana and Varela speak of "structural coupling" instead of adaptation. Organisms and environments are adapted based on structures that somehow "fit" together. Evolution is, therefore, a "structural drift" of organism and environment, each influencing and conditioning the other. This does not preclude, however, that the environment is always more complex than the system, and it is this environmental overcomplexity, that is, contingency or the fact that the environment can change in unforeseeable ways, which poses the problem that systemic order solves.

In the case of meaning systems, the system/environment difference, as we saw earlier, must fall within the system, for the environment must also have some meaning. It is, therefore, difficult to see how society as a whole can fulfill the requirements of self-reference and operational and informational closure. Self-reference requires other-reference.

Operations of the meaning system make meaning, and since meaning extends to include the environment, the system cannot refer its operations to itself, for there is no other from which it can distinguish itself. As we remarked earlier, the world can be the subject of a metaphysical self-reference, but of course, this is paradoxical in that no "other" can be distinguished from the self. If there can be no self-reference without reference to an "other," then we are faced with the metaphysical question of Being. Nothingness "is" in some sense, too.

Furthermore, since the environment must also be meaningful, information is constructed within the system and the environment. We must conclude that for meaning, self-reference is paradoxical. Where does meaning stop and non-meaning begin? Where is the boundary between meaning and non-meaning? How can meaning be a system characterized by operational and informational closure? If not, there can be no constitutive system/environment difference. That from which the meaning system must differentiate itself is always already within it. What is outside society? As long as society was understood to be a kind of thing, a substance, with visible territorial boundaries, it could be observed from without, and there could be something "outside," for example, outside the city walls or the borders, whether cultural or physical. One could stand, for instance, on one side of the border between Germany and France and observe French society. Or one could stand within the city walls and look beyond into a world of uncivilized nature. As long as society was a territory, there was something outside of society. The entire discipline of cultural anthropology and ethnology is built on this assumption, and as the methodological discussion in these disciplines finally realized, the "others" are us.[127] The moment society becomes a system of meaning, it has no boundaries. Communication does not stop at any borders, no matter how heavily guarded they may be. It is a general principle of systems theory that systems are constituted by inclusion and exclusion. What does communication exclude? Of course, within society, exclusions of all kinds are the rule and not the exception. But as a communication system, one cannot even

[127] See Krieger (1991) for discussion of the methodological debate in ethnology.

argue that society excludes non-communication. As Watzlawick noted, one cannot not communicate.[128] It could be argued that since society is a meaning system, it excludes everything meaningless. However, the binary distinction of meaning/non-meaning is paradoxical since even non-meaning must have some meaning to be used by the meaning system as a boundary. Even the distinction between information and noise must be information and not noise since it is a difference that makes a difference.

Social systems, understood as meaning systems, represent a very unusual system. They are all-encompassing, thus violating the principle of selection, which demands exclusion. They are not operationally closed since self-reference depends on distinction from something other, and there is no other. They are not informationally closed since no undifferentiated perturbations are coming from outside that could serve as the stuff out of which information is constructed. Could it be that meaning is not a systemic phenomenon but must be modeled by other principles of order than selection, relationing, and steering? Could it be that although the functional subsystems can be modeled as systems, the society they are subsystems of is not itself a system?

2. *There is no Evolution of Meaning*

The dominance of biological models in systems theory has led to the widespread use of the concept of evolution and its application to systems of all kinds, not just living systems. Physicists speak of the adaptive behavior of complex dynamic systems and the evolution of the universe, and sociologists of the evolution of culture and society. Indeed, it would seem that everything is in some way evolving. But what exactly is meant by evolution? The principles of evolution are variation and selection. We wish to argue here that these principles

[128] Watzlawick (1967). Luhmann on the contrary holds on to "The definiteness of the external boundary (= the distinguishability of communication and noncommunication)" which he claims "enables the operational closure of the world society system" (2012, 87). However, this is difficult to reconcile with the statement that "World society is the occurrence of world in communication" (87), since it is impossible to imagine where the world stops and non-world begins, and it is equally impossible to say it.

are to be found exclusively on the biological level of emergent order and not on the physical level or the level of meaning. We argue that only living systems are subject to the forces of variation and selection. What does evolution mean on the level of life? Living systems are proposed solutions (variations) to the problem of autopoiesis within the constraints of an ever-changing environment (selection). If the problem is not adequately solved or an existing solution becomes inadequate, the organism is "selected" against and it disappears. This is a fundamentally different concept of selection than self-organization, which selects elements, relates them, and steers system operations. As long as a living system can continue its autopoiesis, it is considered to be "adapted" to the environment and thus "viable." As long as the environment is considered "nature" one speaks of "natural selection." It is important to note that no living system selects itself in the evolutionary sense of the term. There can be no *self-selection* in biology or evolution. Systems self-organize and operate to maintain autopoiesis but do not themselves select for viability. Self-organization is not selection in the evolutionary sense. Since the emergence of meaning, nature as well as life have become "integrated" into meaning; it is becoming increasingly "culture" rather than "nature" that selects for or against living systems. Indeed, the much-discussed Anthropocene expresses precisely this fact, implying that contemporary debates about climate change are debates about society and not nature.

What does evolution mean for society? Luhmann argues that such things as physical reality, chemical and organic processes, and even neurophysical and cognitive phenomena constitute the environment of society.[129] We have had occasion to cite Luhmann's remarkable exclusion of human beings from society more than once. But this claim disregards the fact that all these apparently non-communicative phenomena are constructed in communication and endowed with meaning. Otherwise,

---

[129] "The thesis of self-production by communication postulates clear boundaries between system and environment. The reproduction of communications from communications takes place as society. All further physical, chemical, organic, neurophysiological, and mental conditions are environmental conditions. Society can substitute for them within the limits of its own operational capabilities" (Luhmann 2012; xiii).

we couldn't discuss them or describe them in any theory. We couldn't take measures to influence them. And we couldn't manipulate them through technology so they do not select against our projects. The fact that supposedly extra-communicative phenomena are constructed by communication means they could be constructed otherwise or not at all. An organism cannot do this. If an organism does not construct information that is "viable," it vanishes. It must adapt to the constraints of the environment or perish. The social system has no environment in the sense that organisms do or even in the same sense in which the functional subsystems of society do. Luhmann is well aware of this problem and attempts to solve it by introducing the concept of "self-selection." He uses this concept to explain the emergence of functional subsystems within society and to explain cultural and social evolution.[130] If the concept of "self-selection" is not merely another word for self-organization and autopoiesis, then it is an oxymoron, a contradiction, because self-selection is no selection at all in the sense of evolutionary and environmental selection. As Wittgenstein pointed out in his famous argument against the possibility of a private language, if one claims that whatever I say is right is right – this would be self-selection – then there is no longer any right or wrong because the very nature of selective constraints is that they are not arbitrary and are external to the system. How can there be evolution in any sense of the term if no external constraints are selecting which forms of system organization are viable and which are not?

In distinction to the social system, which excludes nothing, the codes of the functional subsystems exclude themselves. The distinctions between profit/non-profit, legal/illegal, certified/not-certified, true/false, and so on cannot code themselves, and they cannot code each other. These distinctions are not themselves either profitable or legal

[130] See Luhmann (1995:190) with regard to subsystems, and (2002:407ff.) for a discussion of social, cultural and political evolution. Luhmann explicitly gives up applying Darwinian evolution to society and claims that natural selection is replaced by building up internal structures reinforced by the systems own feedback. But building up internal structures is nothing other than self-organization and stabilizing autopoiesis. This makes it impossible for the theory of social systems to use the term evolution.

or true or certified; they make it possible to decide what is profitable or not, legal or not, certified or not, within the domain of the system from which these decisions are made. The codes of the functional subsystems construct the borders and limits of the subsystems, and because these are system boundaries, they do not fall within these subsystems. Also, the subsystems effectively ban all other forms of communication than those their binary distinctions can code. The functional subsystems do not allow information to come from outside system boundaries. If the legal system bans certain forms of environmental pollution or racial discrimination in employment, this is not information for the economic system but perturbations to which the economic system must react within the abilities of its own information construction. It reacts by pushing costs onto the consumer or developing business models for pollution control. The functional subsystems can *function* precisely because they are operationally and informationally closed systems. They are confronted by perturbations from an environment consisting of communications that are coded differently, that do not make "sense" within their constitutive borders, and to which they must adapt by means of internal information construction and system conform operations. To speak of a social system beyond the functional subsystems implies that these are somehow integrated into a social whole, a supersystem. How is this possible?

It would seem that the biological and evolutionary idea of environment and selection by the environment does not apply on the level of emergent order of meaning. Therefore, it appears that functional differentiation, which makes society as a whole the environment for the various functional subsystems, at the same time makes "ecological communication" impossible. Once society has become functionally differentiated, no unified and encompassing society can contain its functional subsystems as a kind of supersystem. Luhmann attempts to solve this problem by appealing to interpenetration. But, as we have argued above, interpenetration is merely another word for communication. Appealing to interpenetration simply assumes there can be communication where there cannot. Other ways of dealing with this problem are equally unsuccessful. For example, if one compares society to an organism consisting of functional organs, such as heart, liver, stomach, lungs, etc., it is highly questionable if the metaphor is applicable. The body's organs are not autopoietic, self-referential, and operationally closed systems, for whom the rest of the body is merely environment. The cells of a multicellular organism are not informationally closed to each other.[131] What kind of integration is needed to solve the theoretical problem of systems that are, on the one hand, constitutively and radically distinguished from the environment and systems that exist within other systems as functional elements? In the case of meaning systems, it can be asked if meaning can be conceived as a system or instead as some kind of order in which certain types of systems can emerge and be maintained. This leads to the question of the unity of this encompassing form of meaning. The answer to this question in modern thought has usually called forth theories of rationality. But as Luhmann (1995:477) admits, "Modern societies' principle of differentiation makes the question of rationality more urgent – and at the same time insoluble."[132]

Another attempt to solve the problem of the unity of the system of meaning under conditions of functional differentiation could refer to the status of semiotic coding. Is semiotic coding, which organizes all communication and meaning, fundamentally systemic or not? Is the semiotic code a system code? Or is it a code for some other kind of order? Luhmann is very clear that language alone is insufficient to order social communication. Language is structured by the equally probable options of acceptance or rejection of communication, of yes or no. Successful communication, however, depends upon acceptance and further communication. This is why the emergence of symbolically

---

[131] Levin (2022) has shown that biological electical networks inform indivdual cells how they are to cooperate in order to build organs or morphological structures. Cells to not do what they want, but what they are told to do.

[132] Luhmann solves the problem by reinterpreting rationality as biological adaptability. "Translated into the language of causality, this idea [rationality, AB/DK] decrees that a system must control its effects on the environment by checking their repercussions upon itself if it wants to behave rationally. A system that controls its environment in the end controls itself" (1995:475). This is also the basis of Levin's (2022) concept of intelligence and cognition which he finds in all living systems. It should be noted that this is a biological definition of rationality that has nothing to do with meaning.

generalized media and functional subsystems becomes necessary; they motivate acceptance and guarantee, as far as possible, the continuing autopoiesis of the communication system. Interestingly, Luhmann often seems not to understand meaning as a system but rather as a medium in which systems emerge, specifically, the psychic and social systems. Both are meaning systems and depend on language and semiotic coding. We have already noted severe problems with this view since it supposes that psychic systems use language but do not communicate. We argued above that this makes no sense and that the distinction between psychic and social systems is not well-founded. But beyond the problem of distinguishing between psychic and social systems, there is the problem that meaning is a level of emergent order and not a system in itself. Could the unity of the social system be achieved by thinking of society as a whole as synonymous with a level of emergent order in which the social subsystems arise?

Following this line of thought, one could argue that meaning is like life. Is there a system of life or only living systems? Life is what all living systems somehow participate in. Therefore, life could be thought of as a kind of medium in which organisms emerge, a medium with its own general principles, such as autopoiesis and evolution. But is life a medium? Luhmann understands a medium as a state of loosely coupled elements. Systems arise as "form" within a medium; they arise as closely coupled elements. From a systems theoretical perspective, form means that self-organizing processes of selection, relationing, and steering construct a system/environment difference that constitutes an autopoietic, self-referential, operationally and informationally closed system. There can be no communication between system and environment, and thus, no communication between form and medium. How could meaning as a medium still allow communication between the social subsystems?

Can one think of psychic and social systems as different kinds of systems in the medium of meaning? The problem here, as with theorizing life as well, lies in the constraints of systems theory itself. Systems theory has no basis for the concept of medium other than environment. Systems emerge in distinction to an environment. Neither life nor meaning can be equated with the environment in which either living systems or systems of meaning emerge. Life and meaning are characteristics of the system and not the environment. Just as it is possible and perhaps necessary to consider all organisms as expressions of life, all linguistic and communicative operations are expressions of meaning. The possibility remains to theorize life and meaning as levels of emergent order, as we have suggested. The difference between life and meaning lies in the specific characteristics of these levels of emergent order. For meaning, the particular characteristic, as opposed to life, is that meaning contains its environment within itself. Meaning is not subject to any natural selection by the environment. Meaning is not exclusively directed toward adaptation. Meaning is not structurally coupled to anything other than itself. We will discuss other differences between life and meaning as different levels of emergent order below. For the moment, it is important to note that unlike living systems, of which there are many, there can be only one meaning system. There are many different organisms, but only one system of meaning exists. Indeed, Luhmann argues that there is only one all-encompassing global social system since all communication can be connected to all other communication without any territorial or other boundaries. The social system, for Luhmann, is world society. It is, therefore, questionable if one can speak of social systems in the plural or only of the one, all-encompassing world system. Just as meaning is world, so is society. These issues and questions will accompany us throughout this book. They point to problems within the systems paradigm and toward solutions that may lie in the network paradigm. But let us not get ahead of our story.

According to Luhmann, each functional subsystem of society can only be an environment for the other subsystems. The problem of overall social communication can, therefore, only be solved according to the model of double contingency, i.e., of structural coupling and coordinated behavior. Understanding "subsystem rationalities" does not, in principle, go beyond the model of structural coupling and consequently does not add anything new to Luhmann's model of society. One subsystem, for example, the legal system, does not communicate with the other subsystems, e.g., the economic system, but generates "perturbations" in the environment of the economic system in such a way that the economic system reacts to them according to its own constitutive coding. This is true of the relations between all subsystems. That this model cannot replicate the emergence of meaning – unless, of course, one presupposes commonly accepted structures of expectation, which is explicitly denied here – seems to be confirmed by Luhmann in that it follows from his theory that communication addressed to the whole of society is impossible in a "polycontextural" society:[133]

> Fundamentally, any attempt to make the unity of the system the object of an operation of the system runs into a paradox; for this operation must exclude and include itself in the process. As long as society was differentiated according to center/periphery or according to a hierarchy, at least positions could be fixed in which it is possible to 'represent' the unity of the system without competition, namely in the center or at the top of the hierarchy. The transition to functional differentiation destroys this possibility by leaving it to many functional systems to represent the unity of society through their respective subsystem/environmental differences, exposing them to competition among themselves for which there is no superior standpoint of super-representation. ... ...the unity of society is then nothing other than this difference of functional systems; it is nothing other than their mutual autonomy and non-substitutability. (1990:216)

Luhmann's diagnosis of communicative disintegration in modern society is valuable as a problem analysis. Undoubtedly, the diversity of societal communication is so complex that it results in the formation of subsystems within society as a whole, quite apart from cultural, ethnic, political, ideological, and moral pluralism, or what generally has been called the postmodern condition. It stands to reason that symbolically generalized media can significantly reduce communicative complexity

---

[133] Luhmann orients himself on Gotthard Günther's theory of polycontexturality https://de.wikipedia.org/wiki/Polykontexturalit%C3%A4tstheorie which attempts to formalize relations between radically different perspectives.

in this situation. Nevertheless, the notion of a subsystem itself has no meaning if the overall system is replaced by it, i.e., if "the unity of society...is nothing but this difference of functional systems." The only "system" left are the many functional subsystems, and society becomes nothing but the sum of their differences. In this case, systems theory offers the same vision of fragmentation and disunity as postmodern deconstruction. The internal differentiation of society documents only the internal disintegration of society and allows complexity to grow to such an extent that it can no longer be managed. The problem of over-complexity, it would seem, is "solved" at the cost of generating an even more severe problem. Differentiation leads to a situation in which there is no longer any social communication as such, i.e., no longer social unity. This is not only a problem for (post)modern society but a problem for the theory, which seems to have lost sight of its constitutive object, the social system.

3. Society as a Whole does not have the Unity of a System

The idea that there are subsystems within systems is problematic for systems theory since it must explain how the relation between system and environment that is constitutively necessary to all systems can be conceptualized if systems are somehow contained within other systems. We saw that systems construct their own elements and relate them in certain ways to reduce complexity and achieve some goal. In the most general sense, it can be said the goal of all systems is to reduce complexity by establishing a certain form of order and then operating such as to preserve this organization and maintain the structures and operations that have emerged. All living systems operate so that they can continue to operate, that is, continue their autopoiesis. One could say that all autopoietic systems have only one goal: to maintain their autopoiesis by operating to attain their specific set points. The elements of the system are necessarily functional. The legs and top of a table, whatever they may look like or whatever material they are made of, function as legs and top, or they are not elements of the table system. However, if there are systems within systems, these cannot be merely functional elements like the top and the legs of a table. As autonomous systems, they must be allowed to pursue their general systemic function without regard to the system they are somehow "within," which attempts to ascribe them to a specific function. Indeed, what does the word "within" mean when it is said that internal differentiation repeats the system/environment difference *within* a system? The autopoiesis of a subsystem must be somehow aligned with the autopoiesis of the larger system whose internal differentiation the subsystem represents. How is this alignment to be understood? One explanation is to claim that subsystems set their goals to align with the goals of the larger system in which they operate. The problem with this solution is that one must explain how the subsystem knows the goals of the larger system when the larger system is environment for the subsystem. There can be no communication between system and environment. If the subsystem cannot know to what it must align its goals, it must rely on variation and selection, that is, structural coupling, to achieve alignment. In this case, the larger system, that is, society as a whole, is merely a source of constraints, indeed, an environment, selecting for or against the operations of any subsystem. Subsystems would have to be modeled as structurally coupled organisms in an environment or "ecosystem." We argued above that there can be no "ecosystem" because there can be no system of everything. But if the unity of society cannot be understood as the closure and unity of a system, what kind of unity is it?

Let us ask again what it means to say that a system is *within* another system. To begin with, let us note that in a certain sense, it cannot be said that a system is "in" an environment since the environment is all that is excluded when the system/environment difference is constructed. From the point of view of the system, outside, there is only environment and no other systems. From the point of view of an observer, the system, for example, the frog, can be understood to be "in" the pond and interact with other organisms. The observer, however, is not the system; the observation does not constitute the operational and informational closure as well as the internal differentiation of the system. How, then, is the unity of systems internally differentiated into subsystems to be conceived? Only an observer can see the unity of the difference of system and environment? But when the unity of the system and the environment is in question, how can this be conceived if not as an ecosystem? One could suppose that society constructs the subsystems of which it is composed as "organic" functions, much like an organism constructs a cardiovascular system, a central nervous system, or a digestive system, or much like the cells of a multicellular organism form tissue, etc. On the biological level, the body's various organs depend upon the organizing principles, the DNA, and morphogenetic patterns of the living system as a whole. For this reason, it can be said that the whole is more than the sum of the parts and, indeed, something entirely different. For this reason, the organs and the individual cells do not go off on their own and become pathogenic. For this reason, it is difficult to conceptualize individual cells or organs as autopoietic, operationally and informationally closed systems in their own right and thus their relation to the organism and each other as structural coupling.[134] The relations between elements of a living system are not similar to the relations of closed systems to an environment. Integrating cells and organs in a unitary living system is not adequately conceptualized as structural coupling. Can the integration of functional subsystems into society be conceptualized in this way? And if structural coupling cannot explain the unity of an organism, how can it explain the unity of society?

Concerning the social system and the functional subsystems that compose it, it must be asked what is the DNA, or the morphogenetic pattern, the "code" of society that determines what functional elements it should construct. For society, there is no analogous code. The semiotic code has nothing against contradictions and absurdities. There is nothing that cannot be said in some way. Consequently, no analogous environmental constraints could select or "force" the construction of any particular function for a system of meaning. On the contrary, an organism must take up energy by eating and acquire oxygen by breathing, and some circulatory system or metabolic processes must see to it that nutrients and oxygen are distributed throughout the body. It

[134] Not even Levin (2022) can maintain this kind of autonomy for individual cells in a multicellular organism since these cells must be informationally open to other cells in order to cooperate and build organs or tissue etc. and follow the morphological pattern of the organism.

must be able to recognize danger and react accordingly. But what does a meaning system have to do other than create meaning? And who or what is to say how this will be done or not done? Of course, everyone knows what makes sense and what doesn't, and everyone quickly points out that what others say and do is meaningless. But this use of the difference between sense and nonsense is part of how we make meaning and not an actual boundary of meaning. Beyond this, what kind of economic system, forms of education, government, science, art, or religion are needed for society to be "viable?" Is a hunter-gatherer society less viable than a modern industrial society? Is one religion or worldview more viable than another? How could we judge? What time scale is required? What criteria apply?

Of course, meaning does not exist independently of the physical and biological constraints from which it emerged. There are fundamental constraints. But what they are and how they may influence the order of society is not given, and as mentioned above, such constraints have no selective function. The force of gravity, it might be objected, constrains physical and biological systems and meaning systems as well. But does it? Has society not found different ways to "defy" gravity? What technologies of the future may bring, we can only speculate. The theory of emergent order implies that higher levels of emergent order *integrate* lower levels and contextualize and condition them, not vice versa. As history demonstrates, society is infinitely variable, and the future is open. Where are the principles of variation and selection that would make evolution for meaning systems possible? The biological analogy seems unconvincing. There can be no social or cultural "evolution" in the technical sense. Although such talk is common and, for the most part, unquestioned, it would seem that it has neither theoretical basis nor empirical justification.

As opposed to organisms, meaning systems and society do not achieve closure through exclusion but through inclusion. Any distinction that could be made between the social system and its environment necessarily falls within the system, that is, within society. Business is not politics, and education is not art, religion, or healthcare. But society is society. Where or what is the environment? Luhmann, as we know, says it is individual psychic systems. But we have attempted above to show that this idea is not tenable. For business, one could say that politics is the environment. Political decisions perturbate business, and businesses must construct information out of these disturbances to continue operating in the new regulatory environment. But what is the environment of society as a whole? Nature, both physical and biological, human individuals, and technologies and artifacts of all kinds Luhmann wishes to locate outside society, but all these, as well as persons, are social constructs. They are not outside of society. They exist "as" (hermeneutical "as"!) what they are in meaning. Ecosystems exist only in meaning since they are constructs of an observer. If society has nothing outside, no environment, it is impossible to describe the operations of the social system as a whole in terms of function, that is, in terms of adaptation to an environment. What complexity, other than its own, does meaning reduce? An organism with an appropriate sensory-motor system, central nervous system, etc., can be judged viable if it constructs information out of environmental perturbations to continue its autopoiesis. If lions hunt zebras instead of butterflies, this is because their principle of organization, their code, allows them to construct appropriate information out of everything that is moving in the environment. If this code functions appropriately, the lions will find zebras, eat them, and survive. Society is not like a lion that lives in a particular environment to which it must adapt or perish. Society is only confronted with moral constraints on what is suitable for communication and what is not. Despite moral concerns, one can talk about anything, and in fact, one does. The autopoiesis of society is the communicative construction of meaning. If this stops, then no one is there to tell the story. If it continues, there is no way of judging whether this or that way of communicating is functionally better or worse than any other. Even if every religion, worldview, and ideology claims they know the answer to the best meaning, there is no way to know or decide whether a particular society is viable except after the fact. Civilizations, as history teaches, rise and fall. Religions and philosophies are founded, flourish, and then disappear. Apart from always premature judgments about history, who can say which culture, worldview, language, religion, nation, or people are more meaningful, more communicable, and therefore more viable than another? A pessimist or perhaps a realist would say that history decides. This is indeed so, but until history makes its judgment, assuming those coming later are in a position to appreciate it, what environmental constraints, conditions, limitations, or resistance does communication have to conform to in order to be viable?

Of course, the entire ecological discussion, with its supposition of unity based on structural coupling, can be interpreted as a question of the viability of a particular kind of society within a particular interpretation of the natural world. Nevertheless, we are not dealing directly with nature but with different interpretations of nature. If nature didn't have its advocates to "speak" for it, there would be no ecological issue. Ecology and what it stands for is a conflict within the domain of meaning and not a conflict between society and an otherwise mute nature. If the outside is inside for systems of meaning, we must conclude that perhaps questions of some kind of environmental adaptation of society to nature are misplaced. These questions are nonetheless important because they show that society cannot be modeled as a system. If society cannot be modeled as a system, then even if society consists of communications, these communications cannot be described as operations of an autopoietic, operationally and informationally closed system. Communications cannot be understood as functional elements or even functional subsystems because communication's only function is to continue to communicate. All it needs to do is to refer one communication to another, one meaning to another, and so on. In this process, communication and meaning do not run into any boundaries other than its own making. History shows us how well society does at blocking communication and how these blockages have always been overcome. We do not live in a global world today because communication has successfully blocked communication. The fact that we do live in a global world indicates that if is a unified whole, this cannot be the unity of a system, for there can be no system of everything.

Part II

The Network Paradigm

Chapter 3

Network Science

3.1 Network Science

Networks have a long history.[135] Recently, what has become known as "network science" or even the "network paradigm" is often considered a branch of general systems theory.[136] Although we disagree with this view, it seems quite plausible at first glance. Let us recall that one of the fundamental principles of systemic order is relationing that is, constructing relations between the elements of a system. If one interprets network science as the study of relations, much as cybernetics can be understood to focus on the principle of steering or control, then network science appears to be a part of general systems theory. Network science can be seen as the study of the relations among the elements in complex systems. The complex system that is being studied is resolved into nodes and links, which represent the connections or relations between elements, whatever they may be, whether proteins or people.[137] Also, similar to general systems theory, there are two major methodological approaches to modeling networks. Network analysis can model the nodes and links quantitatively or qualitatively; of course, there are mixtures of the two. The quantitative and, therefore, mathematical modeling of networks is dominant in the natural sciences, whereas qualitative network analysis is typical in the social sciences.[138]

---

[135] For a detailed timeline going back to 1736 when Euler solved the problem of the bridges of Königsberg see Lewis (2009:3-4). It is interesting to note that although network science often speaks of systems, it shares practically no common history with general systems theory and cybernetics.

[136] See (Lewis 2009:7) for a typical statement of this view. "...network science is essentially the science of systems ... networks often model complex systems..."

[137] For an overview of what kinds of topics can be investigated from the network perspective see the collection of publications as https://barabasi.com/publications. A collection of resources for network science can be found at http://www.network-science.org/.

[138] See Hidalgo (2016) for an overview.

As Cerquite and Shi (2022:1) remark in a recent collection of essays on network science:

> In general, networks are mathematical models that describe unified frameworks composed of disaggregated components—the nodes—by including components' interconnections. Network models are then a step beyond the simple case of systems collecting different unrelated entities; indeed, the links between the nodes provide unique information on the investigated phenomena.

The qualitative form of network science is usually represented by Social Network Analysis (SNA). SNA studies the communicative relations among persons in an organization or defined group.[139] Generally, network science, like systems science, is an interdisciplinary field of study, including the social sciences and physics, biology, computer science, and many related disciplines. Lewis (2009:6) summarizes this situation succinctly:

> each subfield using network science had a different working definition. Communication engineers think of networks as systems of routers and switches; sociologists think of networks as influence diagrams representing the social interactions among humans; marketing business people think of networks as populations of buyers; and the physicist thinks of networks as models of phase transition, magnetism, and so on. Biologists use the network metaphor to understand epidemics, genetics, and metabolic systems within cells, and power engineers think of electrical power grids. Network science appears to be in the eye of the beholder with different nomenclature, different vocabulary, and different methods of analysis in each field.

Despite claims to universality and interdisciplinary scope, it can be said that network science, like systems theory, is characterized by a division between the natural and the social sciences, that is, sciences of phys-

[139] For an overview see Wikipedia https://en.wikipedia.org/wiki/Social_network_analysis.

ical and biological phenomena and sciences of meaning. Furthermore, because of the different disciplinary and methodological perspectives, a specific terminology has arisen, which is found in network science and does not depend on concepts derived from systems theory. Nonetheless, some shared concepts exist between general systems theory and network science. Among the most common are "complexity," "emergence," and "evolution." Despite using these terms from systems theory and occasionally referring to networks as systems, network science can do quite well without them as they seem to have no well-defined and well-founded usage in network theory. With the exception of actor-network theory, which we will discuss below, network science does not usually refer to networks as agents. Although self-organization is sometimes applied to networks, for example, in self-organized criticality theory,[140] self-organization does not imply agency. In social science network theory, the nodes of the network are indeed often people who are agents, but the network is not an agent. Network science, except for actor-network theory, as we shall see, does not inherit pan-agentialism, that is, the view that everything is an agent of some kind, typical of general systems theory. Pan-agentialism is an important shared view between general systems theory and actor-network theory.

When speaking of networks, one often speaks also of *graphs*. A graph is usually a visual and mathematical representation of a network. The two terms are, for the most part, interchangeable.[141] Instead of elements, as in systems theory, networks are made up of *nodes* (also called vortices or actors). The nodes are connected by *links* (also called edges, ties, relations, or associations). The nodes (vortices, actors) are connected in various ways and by various amounts (the number of links connected to a node is called a *degree*) by the links (edges, relations, ties, associations) that are described by network modeling. Like systems, networks are

---

[140] See Wikipedia https://en.wikipedia.org/wiki/Self-organized_criticality.

[141] For this reason, Steven Wolfram's (https://writings.stephenwolfram.com/2020/04/finally-we-may-have-a-path-to-the-fundamental-theory-of-physics-and-its-beautiful/) multi-hypergraph model of reality can be considered a network model.

not static structures. They are dynamic and changing.[142] The dynamics of networks can be described on two levels: on the level of "micro rules" governing linking mechanisms and on the "macro" level in which emergent properties and behaviors of the network become apparent.[143] Nodes can appear in various topographical configurations, sometimes called *clusters* or *communities*. There are also "scale-free" networks which contain *hubs*. A hub is a node often said to arise due to "preferential attachment," which leads to one particular node becoming connected to more nodes than other nodes share. Hubs are network nodes with many more links than other nodes.[144] In the social sciences, such hubs are often known as "intermediaries," nodes that have great influence since actors must go through them to link up with other actors in the network. Actor-network theory speaks of all actors as potential "mediators" and hubs as "obligatory points of passage."[145]

It is important to note that network theory is not based as is systems theory on a constitutive difference between system and environment and, therefore, not on operational and informational closure. Networks are not closed systems. Instead, networks are constituted by relations, connections, and openness to changes in configuration and extensions of relations. The difference between systems and networks becomes immediately apparent in the illustration below.

---

142 This aspect of networks is usually where talk of "evolution" comes into network science, but the term should not be taken as a reference to variation and selection as in systems theory but simply a word for the fact that networks grow and change. As mentioned above, change does not imply agency.

143 As Lewis (2009:6) puts it: "behavior of nodes and links is defined by a set of *microrules* governing the behavior of nodes and links. These rules are given at the microlevel, to distinguish them from macrolevel behaviors of networks. Specifically, microlevel rules dictate the behavior of links and nodes, and macrolevel rules dictate the emergence of global properties of a network."

144 See Barabasi (2014) for the concept of "preferential attachment" as well as an introduction to network science and its terminology.

145 See Callon (1986). We will discuss the terminology of actor-network theory in relation to network science in detail below.

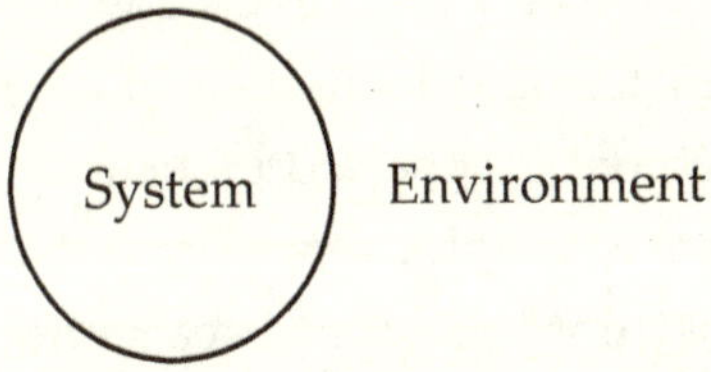

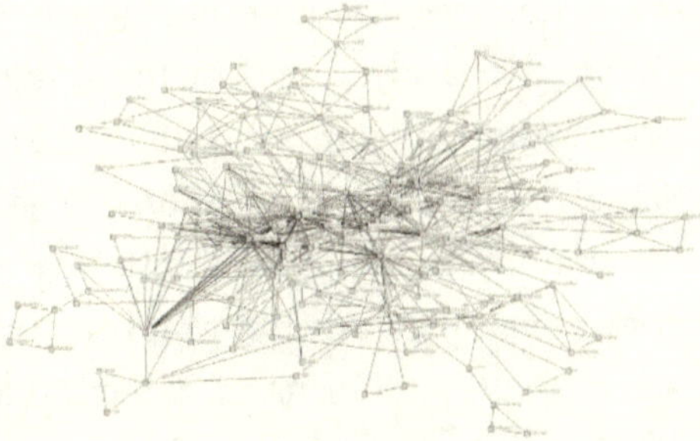

Network[146]

It should be evident from this illustration that we are dealing with two entirely different things when speaking of systems and networks. Nonetheless, a network could be found within a system, as with cell protein replication networks. In this case, the network is a functional element – not a subsystem – of a system. On the other hand, there can be systems within networks, as different companies within a supply chain network are each individual closed organizations or systems linked together by open and changing relations. We will argue that Luhmann's functional social subsystems should be understood this way. Concerning society as a whole, Luhmann's social subsystems are indeed systems but not within a larger social system. Society is an open network, but the functional subsystems can still be modeled as closed systems. The economic system, for example, functions within the open network of society as a whole. We will return to the differences between network theory and systems theory below. For the moment, we are concerned with a general description of network theory, which can be the basis for comparing the two paradigms.

As noted above, there are two basic approaches to describing networks: quantitative and qualitative. Among the two approaches, the quantitative approach abstracts from *what* is connected, the nodes, and focuses on the numerical aspects of the relations between the nodes. It is the intensity, distribution, topology, and regularities of the links that are the object of study, regardless of what the nodes may be. Typical of the

146 Image from arronparecki cc by 2.0 https://openverse.org/image/ca55d60a-f05f-4a41-8e71-582d405d3e8d?q=network

natural sciences, the goal is objectifiable, general regularities and not individual characteristics of the things related. Quantitative network theory abstracts from individual characteristics and focuses on quantifiable, generalizable relations. The method is, therefore, usually mathematical modeling and topological, that is, graphic representation. Typical examples are the graph of Internet traffic or the spread of a disease. On the other hand, the qualitative approach is also concerned with the relations among nodes in networks but derives them from the nature of the connected nodes. The question is not primarily how links are distributed but what the network is made of. The focus is on whether one is studying people, emails, organizations, political parties, etc. Each type of node makes a difference. Depending on what the nodes are, the relations have different meanings. It is not simply how everything in the network is connected; it is a question of what is being connected and why.

Granted these two perspectives, it can nonetheless be said that just as systems theory sees systems everywhere, network theory also finds networks everywhere. After all, what is not a composite of some kind with components connected in some way? As Barabasi (2014:5) notes:

> Networks are also at the heart of some of the most revolutionary technologies of the 21st Century, empowering everything from Google to Facebook, CISCO, and Twitter. At the end, networks permeate science, technology, business, and nature to a much higher degree than it may be evident upon a casual inspection. Consequently, we will never understand complex systems unless we develop a deep understanding of the networks behind them.

The graph, or the representation of a network, can be used to model almost everything from physical processes to protein networks to social networks. Just as systems theory, network theory also claims universality; it claims that network structures are the basis of all phenomena.[147]

[147] See Lewis (2009:8) "...advocates of network science began to demonstrate the universality of network science as they applied it to diverse fields that seemingly had no relations to one another."

Examples of networks are electrical power grids, online social networks, neural networks, both biological and electrical, telecommunication networks, transportation networks, conglomerates, Internet traffic and the network effects of monopolies, markets, economic sectors and regions, protein interaction networks, food networks, epidemics, industry and technology networks, small world networks, etc. When networks are visualized, what is seen is a pattern that is usually a highly complex branching structure but not random. When the methodology is quantitative, what is seen is usually networks that follow a power law distribution explained by "preferential attachment."[148] It is also seen that networks in various areas, whether it be physics, biology, or sociology, exhibit specific common characteristics. This justifies talking about a "network paradigm," which claims that network order is to be found in almost all phenomena no matter what level of emergent order.[149]

Since our concern in this book is the universality claim of the systems paradigm and the network paradigm regarding self-explanation, we will focus on the question of meaning. Can meaning be successfully modeled as a network? Can network science explain itself? Questions such as these imply that the purely quantitative approach will not be the center of our interest. We will not primarily be interested in the distribution or topography of network links but in how meaning conditions both nodes and links. As a theory of meaning, network theory cannot abstract from what is being related, from what it means to be an actor in a network. Our guiding questions will be: What makes nodes and links in a network meaningful? Why does meaning arise from associating nodes with each other via links? The question of meaning is not primarily quantitative unless the amount of links affects what and how the network constructs meaning. We are interested in the meaning, not how many connections a word, idea, artifact, etc., may have with

---

[148] Preferential attachment means that nodes which have more links tend to acquire more links and become hubs. Barabasi (2014) has influentially proposed this model and demonstrated it in many typs of networks.

[149] As an example, see https://www.sciencedirect.com/topics/computer-science/network-paradigm.

other associated meaningful elements. Nevertheless, in some cases, as we shall see when discussing actor-network theory, the number of links can condition meaning. But what is primarily important for a theory of meaning is what a "node" is and means and not solely the number of connections it may have to other nodes.[150] Our question is to what extent network science offers a coherent theory of meaning.

3.2 Actor-Network Theory

The network science literature in the natural and social sciences offers no theory of meaning. It is assumed that one knows what one is talking about, mathematically modeling and describing. The meaning of what constitutes a network from quantitative and qualitative perspectives is simply taken for granted and not considered a question that network science must address. However, this is not the case for actor-network theory (ANT). We admit that the canonical literature of network science usually does not contain references to actor-network theory. Actor-network theory stands outside the usual parameters of what is considered network science. This could be attributed to the fact that although ANT shares basic concepts with network science, it has very different methods and fundamental assumptions about the nature of what is being studied. ANT is concerned with networks from an entirely different perspective than is "normal" network science. ANT was developed in the sixties and seventies, primarily by Bruno Latour. Other significant contributors are John Law, Michel Callon, Steve Woolgar, Trever Pinch, Wiebke Bijker, and others within the broad area of the sociology of knowledge and what has come to be called "science and technology studies" (STS).[151] The unique perspective taken by STS and ANT developed out of questioning such fundamental distinctions

---

150 For an example of the quantitative approach to literature see the network analysis of Shakespeare's dramas. https://www.martingrandjean.ch/network-visualization-shakespeare/.

151 For an overview see Wikipedia https://en.wikipedia.org/wiki/Science_and_technology_studies. For an overview of ANT see Wikipedia https://en.wikipedia.org/wiki/Actor%E2%80%93network_theory. And for an introduction from Latour himself see Latour (2005).

as those between subject and object and society and nature. In addition, ANT used the descriptive methods of ethnology to show that science was deeply embedded in society and that technological artifacts were not passive instruments but more like social partners in the construction of knowledge. The origins of ANT in the ethnology of science explain why all networks are social for ANT. In opposition to systems theory and network science, ANT is not a theory of physical or biological networks but a theory of how physics, biology, and all sciences and their various objects are fundamentally social phenomena. ANT, therefore, is a theory of order on the emergent level of meaning and views all forms of order, including physical and biological order as forms of meaning, that is, as social phenomena.

For systems theory, the world consists of systems. Whenever order appears in the world, it is systemic. Similarly, for network science, order on all levels, physical, biological, and semiotic, exhibits network characteristics. In short, reality consists of networks. This universalistic approach is also true for ANT. Everything is an "actor-network." Actor-networks, however, are distinguished not only from systems but also from the networks that network science studies by ANT's foundational assumptions and methodology.

The first of these assumptions and methodological principles is what could be termed *relational ontology*. Networks for ANT consist of actors and their relations or "associations." In a similar way to how the actors in a novel reveal what roles they play and who they are only in the course of the unfolding story, an actor-network is fundamentally a narrative form of order in which both humans and nonhumans can become actors who play specific roles for certain purposes.[152] Just as in ethnological fieldwork, one looks at how actors are related and then derives what the actors are from their relations. Relational ontology, however, goes even further by assuming that actors are themselves networks. There is no "individual" in the literal sense of the word as something that cannot be

[152] ANT speaks of "actants" which is a term derived from semiotics referring to any role played by an actor in a narrative whether it be human or nonhuman. In a medieval tale an actant can be a sword, the knights horse, a chalice, a castle, or whatever can change the course of events in the story.

further divided, as a "substance," or something that could exist without relations. ANT presumes there are no substances, atoms, individuals, or last elements into which everything could be resolved. In this respect, and in its own way, ANT shares systems theory's insistence that the elements of a system are always constructed by the system within a system and do not exist independently as things in the environment. Reality consists of networks nested within networks. This implies a kind of fractal structure of reality that is fundamentally at odds with the view of normal network science in which networks are first composed of individual nodes that are then linked in various ways. We will have occasion to return to the idea of relational ontology below.

A second important principle of ANT is what could be called *agnosticism*. ANT makes no presumptions about the nature of the actors since it is their associations that determine what they are. Agnosticism is a methodological principle of ANT that says one does not assume anything about what is associated in a network. As is often claimed, one must "follow the actors" without presuppositions (Latour 1987; 2005) to discover the network and who or what the actors are. The actors reveal themselves to the observer by tracing the activities that construct the network. No traditional boundaries or distinctions should be assumed as given. Actors move freely between social domains and even between nature and society.

A third principle of ANT is that all actors should be treated alike; that is, there is "symmetry" regarding whether actors are humans or nonhumans. ANT refers to this principle as "generalized symmetry."[153] Generalized symmetry requires that there are no a priori distinctions between actors. Generalized symmetry has several important corollary principles that characterize ANT. The fourth principle is one of these. It can be called *heterogeneity*. Heterogeneity means that networks are usually composed of many different kinds of actors. An actor-network is composed not only of humans but of things, technologies, institutions, organizations, legal procedures, documents, economic interests, markets, territories, religious traditions, languages, etc. Networks

[153] For explanations of generalized symmetry, see Latour (1996; 2005).

do not respect ontological domains. They associate entities across all sectors. For this reason, it can be said that actor-networks are always heterogeneous and hybrid.

Fifth, and perhaps the most important implication of generalized symmetry is *nonhuman agency*. This principle means that there can be nonhumans in a network and that these material things possess agency and can change the course of a network. Generalized symmetry and heterogeneity imply distributed agency. Technologies, for example, are not assumed to be mere tools, instruments, or passive objects. Instead, they influence the construction of actor-networks, possess their own forms of agency, often called "affordances," and must be considered social partners. Already, general systems theory introduced the idea that order is processual and dynamic. ANT carries this notion a step further by distributing agency across all boundaries and not limiting it to functionalism. Agency, for ANT, is not equivalent to function or the operations of a system. It is a much broader concept that includes any form of construction of associations, that is, networking.

This leads to the sixth principle characteristic of ANT, namely, the priority of *process* over structure. Networks, like systems, are dynamic. Networks are not static structures but dynamic processes of networking. Indeed, one should probably not speak of networks at all but of networks networking. The word "network" is primarily a verb and not a noun. The processual nature of networking, the consequent displacement of structure, and the accompanying goals of maintaining structure typical of systems make networks flexible, contingent, continually changing, and unstable. Not only do networks change constantly, but when they change, the actors also change. ANT's relational ontology prevents actors from being understood as substances with fixed identities. When relations change, human and nonhuman actors change roles and identities.

Finally, ANT fundamentally differs from systems theory by assuming only one kind of network exists. General systems theory assumes the existence of physical, biological, and social forms of systemic order. On the contrary, ANT assumes that all forms of order are actor-networks constructed with "society." However, for ANT, society is not what it is usually understood to be. Society is not one ontological domain in distinction to the domain of nature. The distinction between society and nature is a construction of typically "modern" actor-networks and, thus, a historical artifact that can change. ANT is not a theory of physics or biology but a theory of society that insists that physics and biology must be understood as social phenomena, as actor-networks. ANT maintains a universality claim by arguing that there is nothing outside of society, that is, outside of actor-networks. This claim is understandable if society is understood as meaning and not merely the particular ways in which certain intelligent apes organize their cooperative activities. Meaning is all-inclusive. There is nothing outside of meaning. From the ANT perspective, this implies that there are only actor-networks.

One reason why ANT is not usually included in network science lies in the fact that ANT has developed its own terminology to describe what networking means and what an actor-network is. It was mentioned above that network science generally refers to various mechanisms of network construction. Scale-free networks, for example, form through preferential attachment. Other networks exhibit clustering mechanisms or various topological formation principles that influence network structure. One speaks of "micro rules" conditioning how nodes are linked together and macro structures that emerge as network properties. On the level of micro rules, ANT uses the concepts of "translation" and "enrollment." On the level of network structures, ANT speaks of "trajectories" or "programs of action." We will explain what these terms mean in the discussion below.

In contrast to general systems theory, ANT and network science do not share the mythical opposition of chaos and order underlying systems theory. Networks do not arise as solutions to the problem of complexity or as islands of order swimming in an ocean of chaotic complexity. For networks, complexity is not a problem but a solution. Networks consequently do not have an environment from which they are constitutively distinguished. There is no network/environment difference.

For network science, the world consists of nodes and links; for ANT, the world is constructed out of actor-networks. ANT, therefore, does not project the mechanisms of biological adaptation and evolution onto the level of meaning. There are no biological metaphors in ANT. In contrast to systems theory, for ANT, meaning cannot be modeled as the emergence of systemic order. Instead, meaning emerges as an actor-network. Network order replaces systemic order. An actor-network is not an autopoietic, operationally and informationally closed system that could be classified as physical, biological, or social. Every actor-network is social. Things are social. Technologies are social. Meaning emerges from the mutual and symmetrical agency of humans and nonhumans. Meaning emerges not as a system on the level of human consciousness, cognition, or social communication but as a *network* that necessarily includes both humans and nonhumans, both things and ideas, the material and ideal, both subject and object, and nature and society. Indeed, as ANT proposes, actor-networks are always hybrid and heterogeneous.

In our discussion of ANT, the concept of meaning plays a central role. We must openly admit that ANT does not see itself as a theory of meaning.[154] Surprisingly, the concepts of meaning, information, and communication do not play significant roles in ANT. Nonetheless, we will attempt to interpret ANT as a theory of meaning, and we will make special use thereby of the concept of *information*. This will allow us to compare Latour and Luhmann as proponents of theories that describe order on the emergent level of meaning. In contrast to Luhmann, however, Latour does not speak of self-organization or emergence and does not consider meaning a particular level of emergent order. Where Latour and Luhmann would undoubtedly agree is that what we are talking about is *society*. But as we shall see, there are significant differences in

---

[154] Latour (2013:266) does discuss the theory of meaning in terms of the mode of existence named PRE. "For every mode, there is a distinct theory of meaning, a particular semiotics. If there is an even slightly general metalanguage, it has to be entrusted to the mode of prepositions [pre]. To define the sense of an existent is to identify what is lacking, what must be added in order to translate it, to take it up again, to grasp it anew, to interpret it. consequently, in this inquiry, trajectory, being, and direction, sense, or meaning, are synonyms." We will return to these ideas below.

what each considers society to be. Returning to the central concept of meaning, neither Luhmann nor Latour is alone in focusing on meaning. Heidegger and Wittgenstein, for example, have also made meaning, language, and communication central concepts of 21st-century thought.

Although Latour does not use the concept of meaning, we shall argue that it is meaning that he is talking about. Latour's aversion to the concept of meaning comes from his critique of deconstructive post-modernism as a theory of "discourse" based on signs and structuralist semiotics. According to postmodernism, structuralism, post-structuralism, and hermeneutics, at least as Latour reads them, all the world is a text. There are only signs. In structuralism, signs are ideal, virtual, and immaterial. Latour rejects this view and asks, where are the things, the artifacts, and the technology of which the world is made? We agree with Latour's criticism of structuralist, post-structuralist, and postmodern deconstructivism and explicitly reject notions of meaning that do not include material reality. We are not forced to accept the postmodern version of semiotics or hermeneutics. Our understanding of meaning is not primarily linguistic or semiotic in the usual sense of these terms but explicitly includes things and their role in constructing meaning.[155] In ANT, the inclusion of things of all kinds and, primarily, artifacts and technology in society as co-constructors of meaning is expressed in terms such as "hybrids," "quasi-objects," and "quasi-subjects." Central ideas of the semiotic and hermeneutic traditions, such as interpretation, text, language, and signs, are not simply ignored in ANT. They are taken up and transformed in terms of "articulation."[156] We will have the occasion to explicate these and other special terms in ANT below.

From the point of view of actor-network theory, there are indeed

[155] ANT is well-known as a major proponent of what has come to called "material culture" or "material semiotics" which emphasize that things have meaning and not merely signs or language. On material semiotics see Law (1919)

[156] "Articulation does not mean 'to be articulate' in the sense of being able to speak fluently, but rather 'to join together' in the sense of constructing a composite entity out of heterogeneous parts. This is why I have chosen this term to define not only the action of joining together but also the result of this action. Each of the composite entities made up of heterogeneous parts will, from now on, be called articulations." (Latour 1999:306).

systems, but these are to be understood as a certain kind of network. The autopoietic, operationally and informationally closed systems Luhmann describes as functional subsystems of society can be understood from ANT's point of view as particular forms of network order. They are functional networks that have become solidified into "black boxes."[157] Black boxes are networks in which the complex relations between the actors have been covered over and reduced or subsumed to fixed functions that result in standard input/output operations. The many internal constitutive links, relations, and associations have become invisible. We perceive only fixed processes of input and output, which seem almost mechanical or automatic. For Latour (2005:37ff) this means that the actors in the network have become "intermediaries" instead of "mediators."

> An intermediary, in my vocabulary, is what transports meaning or force without transformation: defining its inputs is enough to define its outputs. For all practical purposes, an intermediary can be taken not only as a black box, but also as a black box counting for one, even if it is internally made of many parts. Mediators, on the other hand, cannot be counted as just one; they might count for one, for nothing, for several, or for infinity. Their input is never a good predictor of their output; their specificity has to be taken into account every time. Mediators transform, translate, distort, and modify the meaning of the elements they are supposed to carry. (Latour 2005:39)[158]

From the perspective of ANT, Luhmann's theory of social systems does not describe fundamental forms of order on the level of meaning, whether psychic or social, but rather black boxes within networks. The challenge of ANT to systems theory is the claim that the concept of "system" is not equated to the idea of "order." Order can also appear as networks; indeed, networks are the primary and original form of

[157] "The word black box is used by cyberneticians whenever a piece of machinery or a set of commands is too complex. In its place they draw a little box about which they need to know nothing but its input and output." (Latour 1987:2-3)

[158] We will return to the all-important concept of "mediation" below when discussion networking.

order, at least on the level of meaning. On the level of meaning, where ANT locates all the sciences and their physical or biological objects of study, one finds actor-networks. Systems are a particular kind of an actor-network, which is why science and technology are embedded within society, or as Latour often says, technology is society.[159] ANT does not tell the age-old and almost ubiquitous story of order emerging from chaos. There are not systems on the one side and chaos on the other, even if it might seem that systems "perceive" the complexity of networks as chaotic and a problem that must be solved. The starting point of both praxis and theory in ANT is more or less chaotic networks. Within these networks, small areas or delimited islands have become rigid, standardized, and functionalized. It is in systems that mediators are turned into intermediaries. It could be said that what is seen as systems are forms of order in which information has been reduced to a high level of redundancy, and the ongoing process of the construction of meaning has become channeled and slowed down, giving the appearance of durability, fixed order, and structure over against volatile agency and its unforeseeable vicissitudes.

In the legal system, for example, redundancy is enforced by state power and the application of the binary code of legal/illegal. In science, redundancy results from applying the binary code true/false by means of established and institutionalized research methods and programs. In each functional subsystem of society, regularities are enforced by the operational and informational closure of the system. These regularities appear to sociologists as structures, institutions, organizations, etc.[160] Durkheim famously referred to these structures as social facts, the specific objects of social science. Luhmann speaks of social systems. From the point of view of actor-network theory, social structures, and social systems are located on a level beneath the originary process of networking which constructs meaning. At the basis of the social in all its

[159] See for example Latour (1991).

[160] Latour does not speak of social systems to describe the black boxes which law, religion, business, politics, etc. appear as, rather, he considers these to be "modes of existence." See Latour (2012) for the theory of modes of existence. We will discuss this text below.

forms – which includes, from Latour's point of view, not only humans but nonhumans as well – is networking and not system building. Theoretically, this is expressed in the fact that for Latour and ANT, the basic principles of order are not processes of selecting, relationing, and steering but processes of "translation" and "enrollment." Translation and enrollment designate the micro-rules that construct meaning by creating associations among actors. In and through associations, actors become an actor-network that has meaning. Meaning emerges not as perception and consciousness or even communication but as the construction of actors in networks of associations. What Luhmann describes as the cognitive and communicative operations of systems of meaning are for ANT operations of translating and enrolling that are performed by both humans and nonhumans. From the point of view of network order, actors can be anything, both human and nonhuman. Therefore, contrary to what is everywhere taken for granted, cognition and communication are not the prerogative of human agents or persons. Let us take a closer look at what these fundamental processes of meaning construction are.

3.3 Translation and Enrollment

The actor-networks that ANT first described were things like daily activities in a scientific laboratory, disputes concerning fishing rights in a small town in France, the development of an electrical railway in France, in general, the social context of scientific discovery and the complex processes by which technologies were adapted and integrated into society. All these empirical studies were done according to the methods of ethnology, and the shared focus of research was always the interaction of humans and nonhumans, especially in science and technology. The classic example of what is meant by nonhumans is technologies, artifacts, but also animals, bacteria, fish, social institutions, laws, and everything else that makes up society. An actor-network is always made up of heterogeneous and hybrid actors. Generalized symmetry and rejection of presuppositions as to who and what can become an actor in a network allows ANT to describe associations in ways inaccessible to traditional sociology and philosophy of science. For Luhmann, technology, artifacts of all kinds, and every material entity must be banished into the social system's environment, which consists only of communications. With this exclusion strategy, Luhmann reflects a long history of opposing society and technology. The typically modern division between society and nature and the material and culture has been expressed in either technological determinism (technology determines society) or in the view that society determines technology (social determinism).[161] Latour rejects both views since, for ANT, technology *is* society. Meaning is always and necessarily material, and the transformation of matter into meaning does not simply add a "virtual" or "ideal" layer onto an otherwise meaningless material world, a level of culture onto nature, but transforms the material world and nature into society. Society, Latour prefers the term "collective," as with Luhmann, is not made up of human beings, but against Luhmann, the "collective" is made up of networks of humans and nonhumans involved in constructing associations which make meaning.

We will attempt to illustrate the theory of meaning that could be grounded in ANT and Latour's work by closely examining a concrete example of an actor-network. Our example will not be any of the actual networks that ANT has described, but an "original," perhaps even the very first actor-network that can be described. Let us go far back in history, even before the appearance of human beings.[162] According to the current state of the archeological record, Homo sapiens appeared

---

[161] For technological and social determinism see the respective Wikipedia articles. https://en.wikipedia.org/wiki/Technological_determinism; https://en.wikipedia.org/wiki/Social_determinism.

[162] This allows us to answer the challenge of Meillassoux (2008) who speaks of the "ancestral" in order to designate a realm beyond the "correlationist" presupposition of modern Western subjectivism, especially since Kant, and as exemplified in Husserl and Phenomenology which influenced Luhmann. The idea of the ancestral questions the assumption that Being – and therefore meaning – is necessarily correlated to human consciousness, to intentionality. Meillassoux asks, how can Being be correlated to human consciousness when there was obviously an "ancestral" world long before human beings arose. Correlationalism is a modern idea arising from the strict distinction between subject and object and the necessity of then somehow putting them together again, since the one cannot exist without the other.

about 300,000 years ago.[163] The first use of tools, the appearance of technology, and all that is implied by it goes back about 3.3 million years. In this period, known as the Paleolithic Age, hominins, our earliest ancestors, used stone, wood, and bone tools for hunting, scavenging, and other purposes. Stone tools can be dated to ca. 3 million years before Homo sapiens appeared. These are the so-called "Mode 1" tools from what is known as the "Oldowan Industry," that is, sites found at the Olduvai Gorge in Tanzania. These tools go back to 2.6 million years. They were made by hammering oval-shaped stones found in riverbeds into forms with sharp edges on one end by breaking away flakes. One end was sharpened, whereas the other end was round and could thus be easily held in the hand. This primitive stone ax, also called the "hand ax," is not the oldest example of stone tools. 3,3 million years ago, stone tools were discovered at Lomekwi 3 in West Turkana, Kenya.[164] This discovery shows that the stone ax is undoubtedly one of the first and even the longest-used tools in protohuman and human history. In many ways, it prefigured later technologies and can be considered a simple model of what it means to use tools or what technology is. When we consider that tool use is often said to be that which distinguishes humans from other animals and thus an essential aspect of culture, the question of this technology could lead to a deeper understanding of human existence than can be attained by beginning with full-blown linguistically mediated conscious experience as does modern philosophy as well as Luhmann's systems theory.[165]

Since the ability to make and use tools and to transmit these skills over many generations is considered a hallmark of culture and civilization, what conclusions can be drawn from the fact that hominins were making and using tools several million years before Homo sapiens appeared?

---

[163] For an overview of the archeological record see Wikipedia https://en.wikipedia.org/wiki/Early_modern_human

[164] See Wikipedia https://en.wikipedia.org/wiki/Stone_tool.

[165] It should be noted that with regard to scientific plausibility Latour's story is much more credible than Parsons' and Luhmann's mythology of the originary encounter of double contingency. See for example the Material Engagement Theory of Malafouris (2013), the Mirror System Hypothesis of Arbib (2012), and Corballis (2003; 2017) on the origins of language.

It implies that human beings, with their big brains and linguistic skills, were not a necessary precondition for the emergence of meaning, or at least, that meaning emerged long before the appearance of specifically human intelligence. Furthermore, it implies that meaning is intimately associated with nature, material reality, and nonhuman forms of life. Let us, therefore, ask: What does it *mean* to use a stone ax? How did this technology emerge? What is the difference between how hominins made and used tools and how behavior that might appear similar in apes at first glance never developed into culture and meaning? Despite all the evidence of tool use among apes and other species, there seems to be some fundamental difference between what animals do and what could be called the non-episodic use of tools on the level of meaning.[166]

Let us attempt a phenomenology of the stone ax. Let us suppose we have some privileged access to the dawn of history and can go back in time and carefully describe, as an ethnologist might do, what happened when the first tool was made, when technology was born, and when meaning emerged more than 3 million years ago in Africa. Let us adopt the methodology of actor-network theory and attempt to discover the original actor-network, which helpfully illustrates all the essential elements of an actor-network and how it can be understood. Of course, this is fiction.[167] But is it not the case that the emergence of systemic order or the emergence of life is also shrouded in mystery? Are not all origin stories, including the event of the Big Bang, in some way or other attempts to illustrate and define first principles with the help of imagination? Be this as it may, what does our phenomenology of the stone ax reveal?

We emphasize at the outset that we are not describing the adaptation

[166] Simian tool use is characterized by its "episodic" nature. See Donald (1991:149), who describes the "culture" of apes as "episodic." "In fact, the word that seems best to epitomize the cognitive culture of apes [...] is the term episodic. Their lives are lived entirely in the present, as a series of concrete episodes and the highest element in their system of memory representation seems to be at the level of event representation."

[167] But fiction is nothing new in science. See for example Luhmann's original situation of double contingency which is just as much fiction as is Hobbes' state of nature.

of an organism to its environment. We are also not offering a phenomenological description of human conscious experience since we find ourselves at a point in time 3 million years before Homo sapiens appears. We see a process by which relations or "associations" are established between particular things and animals so that a unique behavior appears that must be ascribed to all of the participants. Let us call these participants actors and note that the behavior resulting from their association cannot be reduced to any of the actors alone. We see the outcome of a unique and unprecedented kind of *co-operation between hominin and thing* in which agency and what could be called "cognitive function" is distributed among all actors.

> We can never understand and infer the nature of the "cognitive function" responsible for the creation and use of a tool without first recognizing that the various processes responsible for the transformation of raw material to tool, as well as the tool itself, actively and reciprocally participate in the co-construction of what counts as "cognitive function." (Malafouris 2013:163).

In the case of the stone ax, it is a "cooperation" between a hominin and a certain type of stone. What the hominin does cannot be adequately understood as a "reaction" to perturbations coming from an environment. There is no operational or informational closure. There is no purely internal construction of information that may or may not be viable. Therefore, we are not talking about an organism's interaction with its environment. What takes place is a unique cooperation between hominin and stone. Concepts derived from systems theory do not adequately describe what happened when a hominin 3 million years ago, for the first time, picks up a particular stone from a riverbed, holds it in the hand in a certain way, wields it with the arm in specific ways, strikes things, etc. In this world, there is neither a system nor an environment. Concepts such as "system" and "environment" carry the weight of a long history of theoretical development and abstraction from experience. Our ethnologist has gone back in time to witness and describe this event, which is not a single event but a historical process lasting thousands of years. Interestingly, the ethnologist does not see systems distinguished from environments. What do they see?

Our ethnologist sees a hominin picking up a particular stone in a riverbed and holding it in various ways in the hand. The stone has a specific shape, weight, size, and consistency. All these characteristics of the stone suggest, offer, nudge, propose, and influence the hominin so that the stone be held in the hand in a certain way and swung by the arm in a particular manner such that an animal could be struck and killed, or an enemy frightened off, a piece of wood split and formed, etc. It would not be untrue to say that the stone "instructs" the hand how to hold and wield it. Other kinds of stones with different characteristics don't do this. Other ways of holding the stone in the hand, swinging the arm, and striking something don't have the effects that this particular stone and particular movements with respect to these particular objects have. It could be said that this special stone, but not only the stone, also the particular anatomy of the hand and arm of the hominin, as well as the characteristics of the animals, enemies, wood, etc., had certain "affordances" that somehow could be associated with each other such that over thousands of years of acting upon one another created something that did not exist before, namely, a "hunter," or a "warrior" who wielded a stone "ax." Although Latour does not use the term "affordance," it is helpful when attempting to describe the agency of nonhuman actors.[168] Gibson (1979:129) writes:

> [A]n affordance is neither an objective property nor a subjective property; or it is both if you like. An affordance cuts across the dichotomy of subjective objective and helps us to understand its inadequacy. It is equally a fact of the environment and a fact of behavior. It is both physical and psychical, yet neither. An affordance points both ways, to the environment and to the observer."

[168] Latour does not use the term affordances for the contribution of humans or nonhumans to an actor-network. Instead, he speaks of "programs of action" or "propositions" (Latour 1999, 309). A proposition is what "an actor offers to other actors" (309). Also, it should be noted that an affordance is not merely a "constraint," but an *active contribution* to the construction of what is going on, of the "meaning" of the event.

Not only did this unique and exceptional cooperation between certain stones, hominins, and other things that were involved change the being, the identity, and the behavior of the hominins, who became hunters or warriors, it also changed the behavior and identity of the stones as well, which were no longer mere stones, laying about on the ground, but "axes." What is seen is best described as a network of actors, all of whom are transformed by being associated in the way they were and thus becoming a network. The stone became a "tool," an artifact, and the hominin became a "tool-maker." The ax is not a mere stone, and a hunter wielding an ax is not a mere hominin. The ax and the hunter mutually enable and condition each other. There would be no hunter without the ax and no ax without the hunter. Together, they, and of course the animals and enemies that experienced what the ax does to them, build a network in which they subsequently exist "as" what they are. To say, as Heidegger does, everything that appears, appears "as" this or that is to say that hominin, stone, etc., appear as actors in a network of certain associations.

An actor-network is a mutual conditioning of human and nonhuman that makes all "actors" into something that none of them was before. It could, therefore, be said that the associations, the relations, determine what and who the actors are. This is a relational ontology because the actors do not exist prior to the associations out of which they emerge transformed. This mutual conditioning, that is, the construction of associations and relations, is where we propose to locate the construction of meaning. Meaning does not emerge in psychic systems of perception and cognition or social systems of communication, as Luhmann proposes. Our hypothesis is that *meaning emerges as associations constructing actor-networks* in which human and nonhuman actors become what they are. Actor-networks are neither subjective nor objective, neither natural and social, neither material nor ideal. They undercut all the traditional distinctions of Western ontology and systems theory. The process by which actor-networks and thus meaning are constructed Latour calls "technical mediation." What happens in technical mediation?

Technical mediation (Latour 1994) is a process of *translating* and *enrolling* both human and nonhuman actors into actor-networks, that is, cooperative *programs of action*. Latour is well aware that the concept of the "technical" is ambiguous; therefore, he is careful to point out that he is using the word in a special sense.

> Technical … designates a very specific type of delegation, of movement, of shifting, that crosses over with entities that have different timing, different properties, different ontologies, and that are made to share the same destiny, thus creating a new actant. (Latour 1994:44).[169]

The word "technical" refers to the fact that human action is always associated and interwoven with nonhumans, material and biological entities, and their affordances. It implies, against Aristotle's famous and influential distinction between *techne* (constructing artifacts) and *episteme* (knowing first principles), that knowing is doing, making, constructing, and constructing is always the construction of meaning. The taken-for-granted distinction between subjective human beings acting upon objective, passive things in the world does not adequately describe what is happening. The phenomenology of the stone ax illustrates this very well. Our hominin did not sit down and think about how to make an ax. Such "things" or "ideas" did not exist 3,3 million years before Plato. We can assume there was neither language nor thinking in the form we know today. Nonetheless, and this is the crucial point to this entire story, there was meaning. The meaning of the ax was nothing other than the ax itself and not some mental state, a cognitive act, a sign, or an idea. It would take millions of years and the construction of a myriad of complex actor-networks before Plato could come up with the notion that ideas were located in some immaterial realm distinct from things and used as blueprints for their construction, whether by divine or human hand. This traditional understanding of "construction" obscures the original experience of technical mediation. Latour (2013:157) writes:

---

[169] Further: "Technical skill is not uniquely possessed by humans and reluctantly granted to nonhumans.
Skills emerge in the zone of transaction, they are properties of the assembly that circulate or are redistributed among human and nonhuman technicians, enabling and authorizing them to act." (Latour 1994, 45)

> To say that something—a scientific fact, a house, a play, an idol, a group—is "constructed" is to say at least three different things that we must manage to get across simultaneously.

First, the activity of constructing is strangely "doubled." Since there is always more than one actor involved in a network, it is unclear from where the activity comes, from only one, from others, and who was first. Was it the stone that first suggested its use as an ax to the hominin? Or was it the hominin who suddenly realized that this particular stone could have a use? Or was it the enemy who forced one to do something to defend oneself, the wood that needed to be split, or the animal that had to be killed and cut up? Agency cannot be isolated and ascribed to only one of the actors. All actors are involved "symmetrically" in the act of construction.

> Every use of the word "construction" thus opens up an enigma as to the author of the construction: when someone acts, others get moving, pass into action. (Latour 2013:158)

Second, the act of constructing does not know in advance what is going to come out at the end. As Latour puts it, the "vector" of the action is uncertain. There is an "oscillation" between various outcomes or directions.

> the arrow can go in either direction: from the constructor to the constructed or vice versa, from the product to the producer, from the creation to the creator. (158)

Latour cites the example of a puppeteer who is guided by the puppet's personality to pull the strings in specific ways. The hominid did not know it would become a hunter, a warrior, or a builder the moment the stone ax was constructed. Presumably, the stone did not plan this outcome either, and the potential enemy was undoubtedly surprised when being attacked by a warrior wielding an ax. With regard to the construction of scientific facts, it is decisive that the experiment is so constructed that it is not merely the ideas, intentions, and desires of the scientist that appear as results, but instead, what the entities under investigation contribute to the outcome. Technical mediation constructs by means of translation and enrollment of the actors in the roles they play within the network, and none of the actors knows what they will become before the act of construction. The hominin did not set out that day 3 million years ago to become a hunter. The stone had no intention of becoming a status symbol or a cult object that mediated not dead animals or split wood but contact with transcendent forces or helped some hominids find a partner. Construction, as with all networking, is open-ended and contingent.

Finally, the third component of the meaning of construction is what Latour refers to as a "value judgment." Whenever something is constructed, it can be judged as either *well-constructed* or poorly constructed.

> To say of a thing that it is constructed is to introduce a value judgment, not only on the origin of the action...but on the quality of the construction... it is not enough for the experimenter to construct facts through artifices; the facts still have to make him a good experimenter, well situated, at the right moment, and so on. Constructed, yes, of course, but is it well constructed? (159)

The hominin and their children probably tried out countless stones, ways of wielding them, and uses that they might be put to, always judging whether it was well done or ill-done. We will argue below that this meaning of technical mediation as construction directly links action to design. The hominid was a *designer* because the construction of meaning is a question of good or bad design, just as Heidegger's carpenter reaches for his hammer to make a good chair or table, and others judge the result of his work accordingly. Ethics, therefore, is built into reality and not added to it as an "ought" that is opposed to what is.[170]

ANT reaches back into a time before the idea of ideas as somehow distinct from things arose, before the distinction between matter and

[170] For a discussion of the ethical implications of ANT see Belliger/Krieger (2021).

mind had become a foundation of Western thought, and before the idea that what is constructed is somehow "artificial" and not "real." For this reason, ANT is often associated with what is called "material culture" or "material semiotics."[171] When discussing systems theory above, we named the principle of order of meaning systems "semiotic coding." From ANT's point of view, semiotic coding is not primarily about syntax, semantics, or signs somehow referring to things but about technical mediation. What technical mediation does is transform the world into meaning, or in other words, it is the process by which meaning emerges as a level of order on its own. Just as autopoiesis is the process by which life emerges from matter, so is technical mediation, translation, and enrollment, the process by which meaning emerges. Before the actor-network of the stone ax was constructed by technical mediation, there was only the potential of meaning. Just as there was a point in time when there were only chemical reactions on Earth and life had not emerged, the potential was there. When that hominin picked up that stone for the first time, meaning had not yet emerged. What our ethnologist witnessed over 3 million years ago was the emergence of meaning. Meaning emerged as both material and semiotic, both subjective – the experience of the hominin – and objective – the transformation of the stone into an ax. If we consider all the things that are in any way involved in actor-networks as "technology," then this is why Latour can say that technology is not merely a tool, a passive instrument, but a partner, a social actor. Indeed, technology *is* society, and society is always technologically mediated.[172]

As a theory of meaning, the concept of technical mediation does not merely claim that human action is mediated in the sense of being influenced by tools, artifacts, or material entities of some kind, or even visual or acoustic signs, as in language. Technical mediation does not mean that people use tools to do things they might otherwise not be able to do. It does not mean

[171] See Wikipedia https://en.wikipedia.org/wiki/Material_culture; on material semiotics see Law (2019).

[172] In his later work Latour (2013) understands technology as a particular "mode of existence" among others such as politics, religion, science, etc. We use the term technology in the sense of technical mediation which, we maintain, applies to all modes of existence as networks.

that tools and technologies extend or augment natural abilities. This is undoubtedly true. The ax did influence the hominin, just as industrial production influences modern society, human self-understanding, and everything else humans do and are. But technological mediation means more than mere influence, augmentation, or extension of what humans can do based on the natural bodies. In our example of the stone ax, technology constructed the very being, the identity, and the existence of the hominin, who eventually – most probably because of technical mediation – became a human being. Once technical mediation emerged, that is, once meaning emerged, the large brain and linguistic abilities that later evolved had a purpose, something that gave certain hominins an evolutionary advantage.[173] The agency of nonhumans is expressed in the concept of "affordance." What affordances and technical mediation suggest must not be misunderstood as technological determinism.

On the contrary, if mediation is not a mere influence upon an actor but the ontological constitution of the actor, then it is the network of associations among humans and nonhumans that constructs the actor. This implies that the actor is not only the product instead of the producer but also that the producer is technical mediation or meaning. Just as Luhmann can say that communication communicates – and it is communication that produces communicators –ANT says that meaning creates meaning and actors are entities that become meaningful through technical mediation. But here again, one must resist the temptation to misunderstand the term technical and fall into technological determinism. On the level of networking, there is nothing outside of technology to be determined by it. Society is technology. As Latour never tires of saying, the network is the actor. The hominin without the ax is not a hunter but a mere ape, and without the hunter, the stone ax is merely a stone.

The central point of technical mediation is that the idea of mediation refers to the associations among the actors that constitute the actors and not to any influencing done by a subject towards an object or vice versa. The associations first make the actors into what they are and give them

[173] For the influence of toolmaking on the evolution of cognitive and linguistic abilities see Orban/Caruana (2014).

the specific roles they play within a network. As counterintuitive as it may be, it is the relations, the associations, that are primary, whereas the identities and roles of the actors emerge from the relations. The theory of meaning we propose claims, therefore, that being is relational, and ontologically, relations are fundamental, not the relata. ANT is not a substance ontology but a relational ontology.[174] To a certain extent, systems theory can also be considered a relational ontology since systems come into being by virtue of processes of selection and relationing and thereby construct their own elements. Indeed, self-organization is defined as the emergence of order out of interactions between elements. Systems, just as networks, are not substances. But even if we assume that for systems theory, relations are real and not merely added onto reality, systems, as opposed to actor-networks, do not primarily consist of relations. For systems theory, the relationing of elements depends upon their selection and the specific purpose or goal of system operations, and all of this depends on the system/environment difference. This is not the case for networks with no constitutive border or purpose.

Networks, as opposed to systems, are nothing but relations, even if these relations are differentiated into what Latour (2013) calls "modes of existence." It is difficult to claim that what our ethnologist saw when observing the hominin becoming involved with the stone ax was pure relations and nothing else. After all, the stone was there in the river bed. It had been there for perhaps millions of years. How can mere relations change all this? How can one see relations where before that had only been stones? One can see stones and hominins, but where are their relations? We do not stumble over relations as we might stumble over stones. Relations are not things. What are they? We propose the term "information" to designate the associations from which actor-networks arise. For our purposes, which are primarily directed toward a theory of meaning, *relations are information*. What is "seen" when we see relations

[174] Latour (2013:162) expresses this idea by distinguishing between "Being-as-Being" and "Being-as-Other." As opposed to being-as-being which rests upon a foundation of substance or sameness, being-as-other must establish "subsistence" or continuity. It "does have to pass through a leap, a hiatus, to obtain its continuity," and he goes on to cite Tarde (163), "difference proceeds by differing."

is information. Relations appear as information.[175]

Just as systemic order requires that the elements out of which a system is constructed are related in a non-random order, network order also requires that information be ordered in specific ways.[176] Technical mediation constructs the associations or relations among the actors, which allow the actors to appear as actors. What does it mean for an actor to appear "as" an actor? It implies that actors play roles in a narratively constructed "story." Information, therefore, is primarily ordered as a narrative.[177] Actors do things in a temporally unfolding trajectory of events, which assign roles to actors and order events in a certain way. One thing happens after another. One actor does something, and other events follow from this. What describes these roles and coordinates them into a narrative can be called a "program of action."[178] A program of action describes what may be called the "trajectory" of the network and applies to humans and nonhumans symmetrically. As Latour (1992:233) puts it, "Parts of a program of action may be delegated to a human, or to a nonhuman." Agency is always distributed and always a narrative construction.

When systems are said to pursue goals or act to maintain set points, this is a narrative describing who does what, when, and why. Although for both systems and actor-networks, agency can be ascribed to individual actors – for example, Luhmann is compelled to ascribe communication to persons – it must not be forgotten that it is the social system or the network that is the actor. Communication communicates. So also for ANT, the network networks. Agency for ANT, as for systems theory,

---

[175] This definition of information goes back to Bateson's definition of information as a difference that makes a difference, but goes beyond Bateson by claiming that things are relations and nothing else, or in other words, information is reality on the level of meaning, and after meaning has emerged there is no other reality. Or, as Heidegger puts it, whatever appears appears "as" this or that particular thing. The hermeneutical "as" signifies the emergence of information.

[176] Latour (2013) refers to these various ways of ordering information as "modes of existence."

[177] For a discussion of narrative in establishing social order see Belliger/Krieger (2016).

[178] See Callon (1991:136). Programs of action, also called "trajectories," are usually formulated as narratives. See Belliger/Krieger (2016) for a discussion of the narrative construction of social order.

is never individual. For systems theory, it is the steering function of systemic order that accounts for goal-directed behavior. For ANT, "Action is a property of associated entities" (Latour 1994:35). In other words, networks network. Networking is an ongoing process that constructs information ordered in a narrative. For ANT, technical mediation is a form of practice, a problem-solving activity, but one must never forget that the actors doing this practice are not exclusively humans.

To explain how processes of translation and enrollment construct information in the unique sense of the term we are proposing, let us briefly return to our hominin and the stone ax. What is unique and special about this particular hominin and this particular stone? To formulate the question somewhat differently: What distinguishes technology from animal uses of "tools?" When animals use tools, they pick up a rock or stick to break open a coconut or forage for food, then drop it and move on. Animal tool use is characterized as "episodic."[179] The stone ax certainly was also left behind occasionally, but something else was "holding" on to both our hominin and a particular stone that made a difference. Paradoxically, it can be said that the moment a hunter wields a stone ax, quite literally, "things have gotten out of hand."[180] In other words, the stone or the stick is no longer a tool for the moment, only for as long as it is held in hand. Even when it is not being used, even when the hand is doing other things, the tool somehow stays with the tool user, or as Heidegger might say, it exists as "ready to hand" even when not in the hand or not even in sight. The tool does not disappear as a tool and become a mere stone again the moment it is no longer being used, as can be assumed to be the case for animal use of tools. Technical mediation changes everything. When the hominin is not holding the stone, the "ax" and what it means to be a "hunter" is holding on to the hominin and the stone. Neither the hominin nor the stone return to what they were before becoming

[179] Simian tool use is characterized by its "episodic" nature. See Donald (1991, 149), who describes the "culture" of apes as "episodic." "In fact, the word that seems best to epitomize the cognitive culture of apes […] is the term episodic. Their lives are lived entirely in the present, as a series of concrete episodes and the highest element in their system of memory representation seems to be at the level of event representation."

[180] See Belliger/Krieger (2016:29ff.).

involved with each other in the actor-network that is the stone ax. Once it becomes a "hunter" or a "warrior" it remains what it has become even when not holding the ax and not hunting or fighting off enemies. On the contrary, the animal drops the stone when its purpose has been achieved. Nothing has changed. Neither the animal nor the rock has become something other than what they were before. Stone and animal do not become an actor-network. They are not transformed into something they were not already. The actor-network fundamentally differs from any single episodic use of a stone or a stick as animals do this. With the emergence of the actor-network, the stone has become *associated* in meaning with the hunter or warrior as the hominin is with it. Technical mediation constructs the relations or associations between hominin and stone such that something emerges that holds on to all actors and integrates them into a network even when the stone is dropped, and the hominin is no longer using it. To say that "things get out of hand" means that they remain and are "held" in the network and do not simply revert to their former mode of being. They take on their own life and remain what they have been transformed into. That which is holding onto the hominin and the stone beyond any episode of actual use is information.[181]

We have already mentioned Heidegger regarding the primary or originary mode of existence of tools or things in general. For Heidegger, humans' primary or original relation to things is that of *practical use* or practical engagement with the world. The terms "primary" and "original" refer to how Dasein relates to the world before reflection abstracts from immediate experience, that is before one begins to think of the world in terms of appearance and reality, substance and accidents, subjects and objects, etc. Everything appears "as" something specific, as this or that particular thing bound up in some praxis. The originary way things appear is not as passive objects of disinterested knowing. Dasein is not primarily a disinterested observer of objects but an engaged and concerned user of tools.[182]

---

[181] As Latour (2013:223) puts it, "'technology' does not designate an object but rather a difference."

[182] What Heidegger calls "care" (*Sorge*) which characterizes Dasein's existence as practical Being-in-the-World seems to reappear in Latour's return to "experience," but also in his notion of "matters of concern."

Only when the tools break or do not work as expected does Dasein step back from engagement and ponder the situation. But here again, practical engagement with the world, the question of fixing the broken tool, guides attention. The attitude of scientific or objective observation and the traditional assumptions about substance and properties, subject and object, knowledge and reality, nature and society, arose very late in the history of meaning. Things first appear not as passive objects with measurable properties or as substances with so-called "primary" and "secondary" qualities but as tools involved in practices or, as Latour would say, programs of action. The hammer, to cite once again Heidegger's famous example from *Being and Time* – an example that is not too far away from our stone ax – is primarily given in its many practical uses, both what it may currently have, for instance, building a table, and also what it could have, and not as a mere object of disinterested knowing. Recalling Luhmann's definition of meaning from Husserl, it is these many uses, practices, and ways of doing things that make up the horizon of references that constitutes meaning. Luhmann interprets these references to refer to what they are used for. The hammer, for example, "refers" to wood, nails, the workshop, etc., but for Luhmann, this happens only as either perception or communication. For ANT, on the contrary, it really happens in the world of things. Meaning lies in doing, participating in an actor-network, and building tables, not merely perceiving or talking about them. It seems that ANT stays much closer to the practical, primary, and original nature of meaning than Luhmann.

Technical mediation is not "technical" in the usual sense of the word, which describes the workings of a machine or its engineering. Latour's notion of technical mediation is a metaphysical concept that describes how meaning as such is constructed. As Heidegger put it, the essence of technology is not itself technical.[183] When ANT speaks of technical mediation, what is being spoken about is not any artifact or machine. It is also not exclusively a matter of tools unless one wants to consider all things as "tools" in some sense. It is what Luhmann would call the oper-

[183] It should be noted that Latour's technological mediation is much different than Heidegger's *Gestell*, which is a systems concept and not a network concept. Heidegger was describing modern Western society much as Luhmann later described it under the dictates of universal functionalism, while Latour is already beyond modernity and concerned with a post-modern world.

ation of a meaning system. For Luhmann, a meaning system operates to construct meaning via perception and communication and nothing else. This is the autopoiesis of meaning. Unlike Luhmann, in ANT, we no longer talk about "observation" or the functionally directed internal information construction of a closed system. As stated above, network order results from what may be called "networking."[184]

Networking is what happens when things seem to "get out of hand." They are taken up into a network of associations that transforms all the actors involved by allowing them to appear on the level of emergent order of meaning. Meaning, therefore, is not a cognitive act that somehow adds semantic significance, signs, or mental representations to things. Meaning does not reside in signs that somehow stand for things. Signs, significance, language, and much more come much later in history than the emergence of meaning, which we are describing with the example of the stone ax. There is nothing that "has" meaning as opposed to things that do not have meaning. Once meaning has emerged, matter and life; indeed everything that exists, appears "as" something meaningful. As soon as one stone became an ax, all the other stones became non-axes or potential axes.[185] Meaning is not something other than the things themselves. Meaning is not something that things "have" or do not have, something with which a cognitive agent somehow endows them. Meaning, or as we prefer to say, information, is what things "are."[186] And it is only because Being is meaning that cognitive agents can arise and that there is an evolutionary advantage to having big brains and special linguistic abilities. As Latour (2013:230) puts it: "technologies have preceded and generated humans" and not the other way around.

With these remarks, we hope to have set the stage for a more detailed discussion of how technical mediation constructs meaning. Technical

[184] See Belliger/Krieger (2016) for a discussion of networking.

[185] As Wittgenstein noted, it is impossible to say just one word alone. To speak one word is to speak a language.

[186] This is the idea behind what has come to be known as material semiotics, or material culture.

mediation consists of processes of "translation" and "enrollment," just as systems consist of selection, relationing, and steering processes. For any tool to be used as a tool in a specific activity, a series of relations has to be established between the actor using the tool and the tool itself, as well as between the tool and what it is supposed to work upon. The stone ax, for example, works, let us say, upon an animal or a piece of wood. These elements, the hand, the stone, and the animal or wood, have to be related to each other in specific ways. Latour speaks of a "detour" that must be taken by someone who wants to work on something using a tool. In the case of the stone ax, the hominin wished to kill an animal or split a piece of wood. Since they couldn't do this very well with their bare hands, they had to take a detour via the stone ax. Latour calls this detour "translation."

> Translation does not mean a shift from one vocabulary to another, from one French word to one English word, for instance, as if the two languages existed independently. Like Michel Serres, I use translation to mean displacement, drift, invention, mediation, the creation of a link that did not exist before and that to some degree modifies two elements or agents. (Latour 1994:32)

Translation is not simply the establishment of a relation between two previously unrelated things. It is a relation that transforms what it relates. It is not merely a matter of placing two existing things into a relationship with each other, whereby the relationship is something added to the things and may, therefore, also be subtracted. A relation that is added onto already existing relata does not change the things it relates. On the contrary, translation is a unique way of creating associations that transform what is being associated. It transforms them into something new and unexpected. The hominin, who before had only their bare hands to work with, is now a being that can kill animals or split wood by wielding an ax. The stone is no longer something lying in a riverbed but has become an ax, a weapon or tool held in the hand of a "hunter" or a "warrior." It is important not to forget that the animal that is killed or the wood that is split also contribute to this activity because only some animals can become game and only some types of wood can be cut with a stone ax, and this only in specific ways which they decide as much as the ax or the wielder of the ax. When the animal and the wood don't "cooperate," nothing happens. What is essential in this account is that agency is distributed. All actors are equally, or "symmetrically," involved in constructing the actor-network that *is* hunting or chopping. It is not the hominin alone who is the hunter hunting or the builder chopping wood. It is the actor-network constituted by the unique associations between hominin, stone, and animal or wood that the translation process has constructed.[187]

Translation brings *enrollment* with it. Symmetrical agency means that all actors in a network have influenced each other such that they become enrolled into an actor-network that has a particular trajectory or *program of action*.[188] When the hominin sets out to defend the camp against enemies, this is a different trajectory and program of action than setting out to cut wood or hunt animals. Everything is different: the hominin, the ax, the way of wielding the ax, the addresses of the action, etc. It is translation that makes the associations among actors and enrollment that determines the purpose, trajectory, and program of action the network follows. According to the network model, meaning is not what Husserl described as the conscious thematization of an intentional object against the horizon of possible references. Meaning emerges through the translation and enrollment of actors into programs of action that are really carried out and not just thought about. It is only on the level of the emergent order of meaning that actors, both human and nonhuman, are constructed with specific roles to play. Heidegger showed how the intentionality that Husserl described is actually Being-in-the-World, a practical engagement with things guided by "caring" (*Sorge*) or "concern" (as Latour will say). The hammer makes the carpenter in Heidegger's workshop as much as the carpenter makes the hammer. From the ANT perspective, the emergence of meaning constructs the actors: the hunter and the ax, the carpenter and the hammer. It is the

---

[187] The reader is reminded of the threefold meaning of "construction" (Latour 2013:151ff.) including distributed agency, uncertainty of outcome, and value judgement.

[188] Latour will use this idea as the basis for distinguishing "modes of existence," or different kinds of networks.

actor-network constructed by translation and enrollment that creates what it means to be a "hunter," a "warrior," or a "builder," and what an "ax" is. The hunter has different goals than the warrior or the builder. They hold different things in their hands and use them differently. The network and the possibilities that it creates did not exist before the work of translation and enrollment, which enabled a program of action to come into being. When an ape picks up a stone and uses it to crack open a coconut, he drops it again, eats the coconut, and moves on. He has not become a hunter or anything other than he was before. The stone used and dropped has not changed and become something it was not already. Nothing is there to hold on to the ape or the rock after the episodic use, or at least not enough to form an actor-network.[189] When an actor-network appears, however, there is something new. Something remains. Something is there to hold on to all actors in the network even when they move on to do other things. Something has come into being that was not there before. What is this something that has come into being that did not exist before? What is holding on to the actors and transforming them into something new that does not disappear when the activity ceases? We propose calling this special and unique something that exists only on the level of emergent order of meaning "information."

3.4 Information

Unlike systems theory, at least in Luhmann's proposal, network science has not explicitly developed a theory of meaning. Network science takes the nature of the nodes and links that are the object of study for granted and is concerned mainly with how many and in which kind of order the nodes are linked. The meaning of what is in the network is never at issue, or at least, it is not explicitly problematized. Regularities, distribution, topography, etc., constitute network science's object. For ANT, the point of departure for network description is much different.

[189] The use of tools by animals is well-documented and exhibits extraordinary similarities to human tool-use. This seems to imply that we are not dealing with a radical discontinuity, but much rather with a case of emergence. In the same way life emerges from matter, there is always continuity despite jumps.

The associations among the actors are also the object of study, but the associations ANT talks about are an exceptional kind of relation. They are relations that do not merely link actors but constitute actors. For network science, the nodes exist whether they are connected or not. The nodes are first things that then secondarily become related through links in some way or another. What the nodes are, whether computers, people, proteins, emails, etc., is already given, and the links among them are being studied. After all, everyone knows what a person is, what an email is, what constitutes an infection in epidemiology, what users are on a social media platform, etc. Network science is not concerned with metaphysical or epistemological questions about how all these things considered nodes in a network came to be what they are. For systems theory, the situation is different. In systems theory, Luhmann must address this issue because he departs from the usual assumptions of systems theory by proposing meaning as a level of emergent order in its own right. Luhmann, and he is to be credited for this, introduces into systems theory a theory of meaning. As we saw, he relies on Husserl and the notion of the thematic intention of an object against a horizon of possible references. For Husserl and Luhmann, this is what meaning is. However, Luhmann goes on to distinguish between how psychic systems process these references as perception and social systems as communication. He proposes to model meaning as an autopoietic, operationally and informationally closed system.

In contrast to network science, but in agreement with systems theory, ANT is also concerned with the constitution of meaning; perhaps "construction" would be a better word. ANT proposes that associations construct actor-networks in which things first appear "as" (recalling Heidegger's hermeneutical "as") what they are. It is the associations among the actors that "translate" and "enroll" the actors in the roles they play in the network. It could seem that this idea has similarities to systems theory. One could argue that in systems theory, the system constructs its own elements. Elements do not exist before being integrated into a system. No tabletops or table-legs are lying around in the environment waiting to be taken up into a table system. The table system constructs tops and legs out of materials of many different kinds to make a table. No lungs, hearts, or skin lie around in the environment, waiting for an organism to integrate into itself. The organism constructs its own elements that did not exist before being constructed by the living system. Also, in ANT, an actor-network constructs the actors it is made of through translation and enrollment. A hominin becomes a "hunter" or "warrior," and a stone becomes an "ax." But as we will see later, an actor-network is not a system. It is not operationally and informationally closed. It has no environment. It is not functionally defined. It does not "adapt" to anything. We will discuss the differences between systems and networks in detail below. For the moment, it is important to ask: If the associations between hominin and stone construct the actor-network, what do we see when we describe an actor-network?

ANT describes the actors and what they do. But what makes actors into actors with a specific program of action are the associations without which there would be no hunter, warrior, or ax. When the stone becomes an ax, it is transformed into something entirely different from a stone. It is taken up into a network of associations determining what it is. However, associations and relations do not seem to have the same nature as things we can pick up or see and describe. We cannot hold a relation in the hand and use it as an ax. Or can we? What do we see when we see an actor-network? What do we do when we are an actor-network? To answer these questions, we propose introducing the term "information." Information can and does mean many things. We wish to propose a new understanding of what information is well aware of all the risks involved in attempting to change the meaning of a word that is already overdetermined in many ways. Nonetheless, we propose to define information as what we "see" and what we "are" when we appear in an actor-network, when an actor-network comes into being. When Husserl says that meaning is the intentional thematizing of something against a horizon of further references, and when Heidegger pointed out that whatever appears or comes into being appears "as" this or that particular entity embedded in a world of associations, ANT says that networking, processes of translating and enrolling construct actor-networks. In short, networking constructs information.[190] Actors, both human and nonhuman, indeed, all beings are information. If we take the information away, we don't see anything and can't do anything either; the "world" disappears, and we rejoin our simian relatives. Let us not forget that our ethnologist, who went back in time to describe the first actor-network is also an actor in a much more complex and extended network of modern science and ethnology that arose much later in history due to the hard work of constructing networks over centuries. Nonetheless, being an ethnologist is no different from being a hunter or a warrior, for which there are still many examples. We propose to think of all of this, from the first hominin who used the first stone ax to ethnologists, physicists, and biologists of today and everything they are concerned with as information. But what is information?

We have argued that processes of translation and enrollment laid the foundation for the evolutionary advantages of big brains and language. These processes arose long before Homo sapiens with big brains and linguistic abilities appeared on the evolutionary scene. Meaning emerges with the emergence of actor-networks and not with the appearance of Homo sapiens. In contrast to Luhmann's theory, we assert that there were actor-networks and meaning long before the psychic or social systems Luhmann describes. In contrast to Husserl and all of modern philosophy and cognitive science, meaning does not depend on the intentional consciousness of empirical or transcendental egos. Systems theory assumes self-organization based on selection, relationing, and steering processes. As we have argued, this model does quite well for physical and biological systems but cannot adequately describe meaning since systems require a constitutive difference from an environment. The moment this difference becomes meaning, it falls within the system and can no longer function as a boundary of the system. For ANT, however, processes of translating and enrolling distribute agency among all actors, both (proto)human and nonhuman. A network

[190] Luhmann would of course agree that a meaning system constructs meaning, but Luhmann and systems theory do not use the term information in this sense. Although he introduces the idea of a meaning system it is perception and communication that are constructed by the system. Information plays a secondary role.

does not need to distinguish itself from anything; it has no constitutive boundaries and can scale up indefinitely. This corresponds more closely to Husserl's original idea of an infinite horizon of possible references and Heidegger's notion of Being-in-the-World. Translation and enrollment, therefore, are not uniquely human activities; they are not mental processes, they are not forms of subjectivity; they do not take place in the brain, whether it be a big brain or a small one; they are neither psychic nor social nor natural since they are basis for all these distinctions. They are processes of meaning construction in which Homo sapiens came to participate in a unique way millions of years after they had already emerged. The implication is that meaning can no longer exclusively be attributed to human individuals or social systems. Meaning, therefore, is not reducible to what we humans may know or communicate but is the condition for the possibility of being human and knowing anything. Because the world is meaningful, we can understand it and technologically transform it, and after millions of years of evolution, even come to talk about it, and not the other way around.[191]

Because the world is meaningful, having big brains and a command of language can prove to be an evolutionary advantage. But once meaning has emerged, evolution no longer plays a role in how meaning develops. We humans do not give meaning to the world. On the contrary, with the emergence of meaning, a "world" appears for the first time, a world in which humans, for the time being at last, have gained a special place. In words that remind one of Heidegger, we could say that it is not we humans who "use" language, but language which "uses" us. Of course, big brains are important. Big brains and linguistic abilities enable networks to be scaled up almost indefinitely, allowing quick movement through uncountable links. Big brains and language make it possible to

---

[191] This marks the departure of Latour's theory from modern subjectivism in the line from Descartes and Kant to Husserl to which Luhmann despite disclaimers is committed, as well as the solution to the problem of "correlationalism" posed by Meillassoux (2008) which he defines as "the idea according to which we only ever have access to the correlation between thinking and being, and never to either term considered apart from the other." This is not a problem unless thinking is the exclusive property of Homo sapiens or being is considered an object known by a subject.

integrate many more links and relations into a network than possible with more limited cognitive abilities or none. Nevertheless, human cognition alone does not account for meaning, nor is it the source of meaning.

As the phenomenology of the stone ax shows, meaning arose long before Homo sapiens. It is, therefore, not *observation* (Luhmann) but *mediation* (Latour) that constructs meaning. It is not the logical operator of negation, which Luhmann supposes is the foundation of meaning – following Spencer Brown, "draw a distinction" – but the construction of relations and associations, in short, networking, that makes meaning. For Luhmann, it is systems that construct information; for ANT, it is information that constructs networks. Again, we could say with Heidegger that after the emergence of meaning, it is not *dasein* alone but every being that exists as an issue for itself. This is not because all beings are somehow endowed with intentional consciousness or the ability to make observations and draw distinctions but because all beings are endowed with agency in the sense of being able to enter into processes of translation and enrollment. All beings, as beings, are potential mediators capable of becoming involved in constructing relations. This is what the emergence of meaning means.

When we speak of information in this context, let us state clearly that we are not speaking of some mental state. Information, as we wish to define it, is not a cognitive act. Information is not, as Luhmann proposes, observation, that is, binary distinctions founded upon negation. Furthermore, according to Luhmann's theory of communication, information is not that which is talked about in communication, one of three selections along with utterance and understanding. Information in the view we are proposing is a form of being *sui generis*. It consists of relations and nothing else. On this level of being, that is, on the level of emergent order of meaning, there are only relations, and what exists does so *because* it is constructed of associations through the mediating activities of both human and nonhuman actors, who themselves can only perform such activities because they are taken up into processes of translating and enrolling.

Emergence means that phenomena appear that cannot be derived from anything other than themselves. Information can only come from information, just as for Luhmann, communication can only come from communication. Again, information, as we define it, is not a special kind of thing. Information is not a substance with attributes. *Information is a process, the process of networking.* We cannot hold information in our hands, on paper, or in a database. We do not *have* information; we *do* it, or rather, it does something with us, and it is through information that we become what we are. Information is not a new name for a particular kind of substance. Information must not be interpreted on the basis of Western metaphysics or on the computational information theory that dominates discussions of the topic today. In the view that we are proposing, it is through the ongoing construction of information by information that humans and nonhumans come to be what they are and do what they do in a meaningful "world." It cannot be emphasized enough that information emerges not as mental states or linguistic acts, consciousness or communication, or the transfer of bits and bytes but as actor-networks. Information is at once material and mental, at once human and nonhuman, at once signified and signifier, at once given and constructed, at once subject and object, at once nature and society, at once word and thing, at once map and territory. Indeed, all these traditional distinctions are not helpful and even misleading when defining information. This situation is not new.

Since the concept came into focus at the beginning of cybernetics, the ontological status of information is notoriously uncertain. Wiener's (1949) famous declaration that information is information and not matter or energy says more about what information is not than about what it is. In addition, Shannon and Weaver's mathematical theory of information has come to dominate computational models and has uncritically been taken over by media theory, communication theory, and other social sciences. The result is that information has become a term that is not very informative. It is, therefore, not surprising that Latour does not use this concept and has not explicitly developed a theory of information.[192] We wish to intervene in this unsatisfactory situation by suggesting an

---

[192] Floridi (2011) is one of the few who explicitly attempts a systematic philosophy of information. For a critique of Floridi's position which overlooks the relational nature of information see Belliger/Krieger (2021)

interpretation of Latour's principle of "irreduction" as a definition of information and as the foundation of a theory of information. The principle of irreduction was enunciated by Latour in an early philosophical work of the same title. In this short essay, which was published together with his study of Pasteur, Latour says that "nothing is, by itself, either reducible or irreducible to anything else" (Latour 1993:158). What does this mean? If something can only exist insofar as it is neither the same, that is reducible to, something, nor totally different from (that is, irreducible to) something, then it can only exist as a relation. In short, nothing can exist "by itself." This implies that Being is relational.

We spoke earlier of ANT's relational ontology. According to a relational ontology, identity, sameness, continuity, and persistence are not given but constructed from relations. What do relations do? Relations *mediate*. Mediation implies that whatever exists, exists because it is a mediating mediator. Based on this principle, it becomes understandable why technical mediation constructs information. It does so because it creates associations, relations, and processes based upon constantly overcoming reductions to identity.[193] If we interpret the principle of irreduction as a definition of information, what it states is that information emerges as the mode of being *sui generis,* which integrates actors into networks and transforms the actors, elevating them, as it were, onto a new level of emergent order, the level of meaning. As with all levels of emergent order, meaning *integrates* the levels below it: the physical and biological. Once meaning has emerged and "world" appears, nothing exists outside or beyond the world. Physical and biological science and all they assert to know exist not outside of meaning but within the world of meaning. They are parts of what Luhmann would call "society." Latour does not speak of society but of the "collective" to emphasize that nonhumans are also involved in constructing meaning. From this perspective, it is understandable why ANT arose out of Science and Technology Studies and the sociology of knowledge. What happens in the scientific laboratories is through and through social, that is, the construction of actor-networks. The fundamental distinction between

[193] See Latour (2013) for a detailed discussion of this aspect of networking.

subject and object, society and nature, facts and opinions, the ideal and material, and many other such dualities that Western modernity has created are placed by this view into question. Indeed, Latour (1993) can even claim that if one looks closely at what is really happening in the world, one sees that "we have never been modern."

Perhaps we must accept that hardly any traditional theories used to define information adequately describe what information is on the level of emergent order of meaning. Again, information is not something we have and could perhaps do without; it is what we are.[194] Following Latour's principle of irreduction, we can claim that what exists is information and nothing else. Where Luhmann proclaims, "There are systems," Latour declares, "There are networks." Both are universal claims. Both talk about meaning and information, but they do so in entirely different ways. Systems are not networks. It makes a difference if information is produced by the autopoietic operations of an operationally and informationally closed system or whether the actor-networks in which we exist are constructed by information. For systems theory, we, that is, the system, construct information, whereas for ANT, it is information that constructs us.[195]

---

194 Floridi (2014) speaks of humans as "inforgs" instead of cyborgs. Inforgs are beings that exist as information in a world that is also information, an "infosphere." Floridi's concept of information, however, is different from what we are here proposing. We are equating information with meaning. It is only very much later in the history of information when under certain circumstances it becomes useful to distinguish information from meaning, for example, when studying animal communication, or developing a mathematical theory of information.

195 In this context it is interesting to note that there is a tension in Levin's (2022) theory of universal agency and collective intelligence which is founded on systems theory. For Levin, individual cells are at once autonomous, operationally and informationally closed systems and still they necessarily obey the instructions of morphological bioelectrical networks. If cells can only cooperate and form collective entities such as organs or limbs by being instructed by a bioelectrical network, then they cannot be autonomous. They do not do what they want, but what they are told to do. They are not informationally closed but informationally open. If all intelligence, as Levin claims, is collective, and collectivity demands informational openness, then intelligence must be a network and not a system.

3.5 Society: System or Network?

We have asked repeatedly throughout this book whether society could not be better modeled as a network than as a system. Coupled with this question was whether or not the universality claims of systems theory or ANT were justified, that is, the claim of the theory to be able to explain everything, including itself. Starting from Luhmann's introduction of a third level of emergent order, the level of meaning, and then quickly focusing on "society" as the form in which meaning is ordered, we noted that Luhmann was very clear in distinguishing his approach to sociology from traditional social theory. Luhmann excludes everything from the social system that is not communication or the way in which meaning is expressed beyond individual psychic experience. From the point of view of systems theory, at least four traditional assumptions about the nature of society must be rejected:

1) that society consists of actual people and relations between people

2) that society is constituted or at least integrated by consensus among human beings, by concordant opinion and complementary purpose

3) that societies are regional, territorially defined entities, so that Brazil as a society differs from Thailand, and the United States from Russia, as does Uruguay from Paraguay

4) that societies, like groups of people and like territories, can be observed from outside. (Luhmann 2012:6)

Luhmann did away with the first assumption that society consists of human beings and their relations by declaring human individuals to belong to the social system's environment, which is a system of communications and not of people. All communications are on the same ontological level, the level of meaning. People, as such, have no meaning other than what is said about them and attributed to them in communication. Although Luhmann distinguishes between society as a whole, large functional subsystems, and small interactions among "persons," this does not carry the traditional distinction between individual and society, agency and structure, and micro and macro levels that characterize traditional sociology. For Luhmann and Latour, social space is "flat."[196] Systems theory does acknowledge social differentiation in the form of systems within systems; indeed, this is what "internal differentiation" in general systems theory means. Nonetheless, if subsystems are indeed systems, then they relate to the systems they are "within" as to an external environment. This means that persons and their face-to-face interactions are only in the social system to the extent that interaction systems construct them, that is, only as communication. There is consequently no problem in sociology of "integrating" individuals into society. There is no fundamental opposition between agency and structure and no micro or macro levels of society that are understood as separate ontological domains.

Social systems theory eliminates the second assumption that society is based on consensus because it does not assume that communication depends on agreement. Indeed, the opposite is the case.[197] The more disagreement and controversy, the more communication, and the more agreement, the less communication, for there is less to talk about. If everyone were to agree upon what counts as true, socially acceptable, and as honest and sincere, as Habermas (1987) claimed was necessary for communication to be possible, then there would be no controversy, no conflict, no diversity of interests, no real politics, indeed, no society as we know it. All the real problems would already be solved. The fact that the forces of social constraint and conformism are so great, a fact that for Luhmann is illustrated by the emergence of symbolically generalized media and functional subsystems, testifies to the constant danger of communication failure and the threat that this poses to the ongoing autopoiesis of the social system. Society is based not on consensus but on the need to reduce communicative complexity. Systems are always a complexity gradient, a balance of internal and external complexity.

---

[196] This means that there are no ontological domains and no micro and macro levels in society. For Latour's version of this idea see (2005:165ff).

[197] Habermas's *Theory of Communicative Action* (1987) argued that communication depends upon a consensus about fundamental criteria of truth, correctness, and sincerity.

Systems self-organize and come into being as attempts to reduce complexity, which they necessarily presuppose, and not based on sameness, conformity, unanimity, and consensus.

The third assumption that societies are territorially bounded entities confuses society, at least in the modern period, with nation-states. Nation-states, ethnic regions, cultures, etc., should not be equated with society. At least in the modern period, where globalization removes any territorial constraints on communication, it must be assumed that nation-states exist within society, the one global society, and are not society itself. If the social system is defined as the system of communications, where are the boundaries of communication? Does communication stop at the border of a nation-state? The theory of social systems as a theory of communication is a theory of one global society, a "world society"[198] consisting of obviously not territorially bounded communication. All the functional subsystems of society are global. Business, science, law, religion, art, education, etc., once differentiated into functional subsystems with exclusive/inclusive binary coding, are not territorially bounded. The only exception is the political system, which has the function of making collectively binding decisions whereby the "collective" is defined as a territorially limited nation-state. The collective of the nation-state, however, is not the collective of global society. There is no political system on the global level, which implies that society currently lacks the political function. How can the territorially bounded nation-state "steer" society without powers beyond its borders? And when nation-states attempt to steer global society, there ensues a Hobbesian war of all against all on the international level.

Finally, assumption four about the possibility of external observation of society is rejected because it would place the observer outside society, which is impossible. After all, the social system is a meaning system, and one cannot go outside of meaning. Meaning alone can observe meaning. Meaning, for Luhmann, exists as self-observation or self-differentiation in perception and communication. The self-reference of the

---

[198] See Luhmann "Globalization or World Society: How to Conceive of Modern Society?" published in International Review of Sociology (1997).

meaning system is a major topic in Luhmann's theory and takes up much of the early discussion of second-order cybernetics. If society is a system of meaning, then sociology must become a theory of meaning, which, as we have argued, implies that it must be able to explain itself and not merely social "objects" that are observed in a domain located somehow outside of the observer.[199]

Having quickly reviewed Luhmann's four objections to traditional sociology, let us turn to ANT. Latour would agree with Luhmann in rejecting these typically modern assumptions about society and the nature of the social. For ANT, just as for Luhmann's theory of social systems, the social and society cannot be reduced to human beings and their interactions. Contrary, however, to Luhmann's strategy of exclusion that bans individuals and everything that is not communication from society, for ANT, the "collective" excludes no one and no thing. Technologies, artifacts, all things material, and living are in some way social partners.[200] Whereas Luhmann's concern is to limit the scope of sociology to the study of forms of communication, for Latour, sociology finally includes all those actors that contribute to and actively participate in the world of meaning. Of course, Latour intends to overcome typically modern opposition between individuals and society. If society cannot consist of human beings, neither does society consist of the hypostasized relations among individuals. Such relations were thought of as "social structures," what Durkheim referred to as "social facts," that exist on a supposed "macro" level above the "micro" level of individual interactions and which then control individual actions and thoughts, as it were, from behind the backs of the individuals. Against traditional sociology, Latour insists that no individuals can somehow exist on their own in an a-social state of nature until they sign a social contract. Then, once they have signed this mythical document, social structure, insti-

[199] Philosophical hermeneutics, Heidegger, Gadamer, Ricoeur, has always claimed that the social sciences in distinction to the explanatory methods of the natural sciences were oriented toward "self-understanding." For a discussion of the once hotly debated issue see K.-O. Apel (1988).

[200] This becomes apparent when Latour (2013) officially includes not only technology, but also material things and living organisms as a "mode of existence" within the collective.

tutions, and organizations arise, which make up what Durkheim saw as the proper object of empirical sociology. For ANT, social space is not divided into distinct domains of agency and structure; it is "flat."[201] The distinction between individual and society is a modern myth disguising the networking that is constructing the collective.

Second, just as for the theory of social systems, for ANT, consensus or agreement cannot be the basis of meaning. On the contrary, meaning is the basis for both consensus and conflict. Relations and associations, which are the very stuff of the social, are not always peaceful and hardly ever reflect agreement about values, world views, or who and what and how things are supposed to be associated and what trajectories networks are to follow. Disagreement is just as meaningful as agreement; conflict is probably more prevalent than accord. For something to be social, there is only the demand that it, whether human or nonhuman, enter into associations. And something that does not enter into associations does not exist. Actor-networks are often conflictual, continually in revision, pursuing different and contrary goals, and open to transformation in all directions. The idea that people can only meaningfully talk to each other if they agree on fundamental assumptions about reality, values, etc., which Habermas proposes, rests upon a very simplified understanding of communication. It is only when people disagree on fundamental values and world views that communication becomes crucial. Otherwise, we all know what is expected of us and act accordingly, social life collapses into a sterile conformism, nothing new and surprising happens, information is not constructed, and the network does not grow.

Thirdly, for ANT, as for the theory of social systems, society cannot be understood as being territorially bounded. First, networks, by nature, do not have clear borders or boundaries, especially not territorial ones. They have distributions, intensities, extensions, branches, hubs, and peripheries, but nothing marks exactly where a network stops and something else begins. The actor-network of our hominid hunter quickly extended to a network of a builder or a warrior. As soon as the

[201] See Latour (2005).

stone ax became a status symbol, it became a network of social order and advantages in finding a mate. The stone ax also became a ritual object in a religious ceremony that served to maintain contact with transcendent forces. The process of extending and transforming the network trajectories continues until today. Furthermore, actor-networks are constructed of information; as a relation, all information is, in principle, linked to all other information. Information cannot be territorially bounded, not even by firewalls. Information cannot constitute a bounded individual or what Western metaphysics has called a substance or an individual. General systems theory also has no use for substances or individuals since systems are, by definition, compositions of elements constructed by the system. Systems theory, however, does require that systems be bounded by the constitutive difference to an environment. We have seen that this theoretical demand causes problems for Luhmann's theory of society as a system of meaning and communication, for meaning is information, and information is relational. The boundary of a system of meaning must itself be meaningful and thus fall within the system, or to put it more precisely, link the outside to the inside such that no real boundary exists. On the level of emergent order of meaning, relations are real, and every boundary is a relation. Therefore, the necessary exclusion of an environment constative of systemic order in all its forms is lacking in meaning systems. But without this constitutive boundary, there is no selecting, relating, and steering of system elements and operations. As argued above, this implies serious problems for a systems model of meaning. Since meaning is all-inclusive, and there can be no system of everything, it would seem that Luhmann's attempt to understand meaning as a form of systemic order should be revised. We have argued above that the unique characteristic of meaning as constituted by information makes it impossible to model meaning as a system successfully. Meaning knows no boundary other than such that it itself constructs for various purposes. We argue that conceiving of meaning as a network and not as a system avoids this problem. Networks are not constituted by boundaries, as are systems, but by associations, links, or relations. Networks are scalable. They can expand almost indefinitely or shrink to small groups of actors and few associations. Networks differentiate, as we shall see below, but never by binary exclusion. For

ANT, meaning is necessarily ordered as a network, and only under certain circumstances, where boundaries can be temporarily drawn to construct functional black-boxes, can it be modeled as a system.

Finally, the fourth assumption that traditional sociology makes when it proceeds from the idea that society can be observed "objectively," that is, from "outside," has been debunked by ANT's close description of how science actually works. Science in action, the title of one of Latour's first works, shows that the traditional epistemology, which understands science as a disinterested knowing of supposed facts, is mostly myth. ANT has for decades documented how many social processes construct scientific facts. Society, no more than anything else, can be an "object" for a" subject" in the sense that modern epistemology and philosophy of science supposes. Not only ANT but Science and Technology Studies have shown that science is not made up of subjects and objects. What things are and how they appear in laboratories results from long and complex constructions, interactions of humans and nonhumans, indeed artificialities. The ideas of the subject and the object portrayed in modern philosophy of science have nothing to do with science as it is practiced and are largely modern myths. What philosophical hermeneutics has always claimed turns out to be true not only for the social sciences but for science in general. All knowledge is interpretation, and there is no dichotomy between "understanding" and "explanation." Different kinds of networks construct information in different ways and thus embody different kinds of knowing. But none of these "modes of existence," as Latour (2013) calls them, are adequately modeled according to the traditional distinction between subject and object and the idea of a pure, value-free, and disinterested knowing.

At this point, we can ask what the rejection of these four traditional principles of modern sociology by both Luhmann and Latour means for the self-understanding of the social sciences. If both Luhmann and Latour mark a clear transition from classical modern social science to what could be called a "post-modern" theory, what is the "object" of the social sciences, and what is the function of science in society? Let's look back at the development of the science of sociology. We see that it has been based on a fundamental assumption of the difference between culture or society on the one side of an ontological divide and nature on the other. Modernity divided reality into ontological domains, with nature on one side, society on the other, the object on one side of an ontological divide, and the knowing subject on the other. This distinction goes back to Descartes' distinction between res cogitans and res extensa, that is, thinking beings who are free and somehow cannot be reduced to matter on the one side and natural beings who exist under the rule of determinate causality on the other. This ontological divide grounded the typically modern self-understanding of the human (Humanism) and the later distinction between the natural sciences and the social sciences. Sociology, as it was conceived at the end of the 19th Century, was a science concerned with free subjects who enter into relations with one another to build society and culture. It was a science of the relation between the one and the many. The fundamental question was how individuals are integrated into society and how society conditions individuals. Although it was admitted that society does not obey the determinate laws of nature, if there is to be a "science" of society and not merely philosophical speculation, it must be assumed that society obeys its specific cultural and historical laws. The latest version of these laws is what network science discovers, as well as the new social physics.[202] These were the laws that Durkheim, for example, claimed to be discoverable by the newly founded empirical science of sociology. Whatever these laws may be, they are what the social sciences have been looking for ever since.

Both systems theory and ANT reject the traditional assumptions of sociology and allow us to go directly to a different conception of society and sociology. Although Luhmann must be credited for introducing meaning as an independent level of emergent order beyond matter and life, a level of order upon which society comes into being as a system of communication, the ontological status of matter and life in relation to meaning in Luhmann's theory remains unclarified. For Luhmann, as for all traditional sociology, the realm of nature, things, biological

---

[202] For an overview see Wikipedia https://en.wikipedia.org/wiki/Social_physics.

bodies, artifacts, and technology are not part of the social system. There can be nothing in the social system except communications. Everything else is banned into the environment, which is assumed to exist, somehow beyond the city walls of the social system. Luhmann carefully maintained the fundamental distinction between society, nature, and even society and technology. For Luhmann, all these things are, in a certain sense, part of the social system, but only insofar as they are the subject matter of communication. Society is communication, not material things, living beings, artifacts, or even human individuals. The social system consists of communication and not what is talked about, except when communication talks about itself, which is a special case whose difficulties have led to our criticisms. The things that are spoken of are not part of the social system; only their meaning in the form of selected information is part of society. When things become information, they can be communicated and, in this sense, have meaning and are part of the social system. But information is not what they are, their mode of existence, as Latour will say. They exist in their own ontological domain, whether or not they are being talked about. For the social systems theory, the typically modern gap between society and nature remains wholly intact and unquestioned. Physical and biological systems do not exist on the same ontological level as meaning. There is meaning and other kinds of systems, whatever that might "mean." For Latour, on the contrary, meaning is constructed not by communication but by networking or, as we have seen, technical mediation. And since there is nothing that does not appear in an actor-network, and thus as in some way or another technically mediated, meaning is not in any sense of the word a "realm" somehow distinct from matter, life, or artifacts. Meaning is not one domain of being alongside others. Meaning is actor-networks, consisting of both (pre)humans and nonhumans. The world, which is the only world we know, is always already "social" because it and everything in it is constructed by relations. And since relations are information, the world of relations is always already meaning.[203] Actor-networks are always at once natural and social. This

[203] We should not forget that the moment one particular stone became an ax, all the other stones became "not-axes" or "potential-axes."

is what Latour's principle of irreduction implies. There is no nature "out there" conceived of as a "thing in itself" that we cannot know – Kant's critics quickly asked how he could know about the noumenon – or a realm of objective facts governed by determinate causality and ontologically distinct from society as a realm of freedom, subjective opinions, cultural artifacts, and historically changing institutions.

Therefore, the sociology that ANT proposes represents a more radical break with the tradition of modernity than Luhmann's systems approach. To mark this fundamental difference between ANT and traditional sociology, Latour prefers not to speak of sociology at all but instead of "associology," a theory of the *associations* of humans and nonhumans that make up the world of meaning. As Latour (Latour 2005:5, 9) puts it:

> Even though most social scientists would prefer to call 'social' a homogeneous thing, it's perfectly acceptable to designate by the same word a trail of associations between heterogeneous elements. Since in both cases the word retains the same origin—from the Latin root socius—it is possible to remain faithful to the original intuitions of the social sciences by redefining sociology not as the 'science of the social,' but as the tracing of associations. In this meaning of the adjective, social does not designate a thing among other things, like a black sheep among other white sheep, but a type of connection between things that are not themselves social.

For Latour, the science of sociology cannot define itself in opposition to or as somehow similar to the natural sciences, nor can the natural sciences continue to assume they are discovering objective facts of nature independent of all social concerns and embeddedness in society.[204] It can be said that ANT extends and completes the universalization program of sociology that began with Spencer and Durkheim and was continued by Luhmann. The idea of the sociology of knowledge

---

[204] Latour (2004) proposes that one speak of "matters of concern" instead of matters of fact. The very idea of an "objective" fact of "disinterested" knowing is an illusion of Western modernity and ignores how knowledge actually comes into being.

completed this program. But ANT goes even further in that it acknowledges that sociology is no longer a science of society distinction from the science of nature or even that such ontological domains as nature and culture exist. Things, artifacts, and nature are no longer excluded from the realm of the "social" understood as everything that is "associated," that is, everything that is networked and thus constituted by information. The traditional distinctions between domains of scientific investigation and disciplines require revision.[205]

Despite the theoretical advances of general systems theory and ANT, the distinctions and assumptions typical of Western modernity remain mostly unquestioned. Although for both Luhmann and Latour, traditional modernity has passed, the convictions of classical modern thought have not lost their persuasiveness and influence. Latour suggests in a provocative and radical statement that we should acknowledge that "we have never been modern" and leave the conundrums of modern philosophy, nature and society, subject and object, idealism and materialism behind as we move into a networked future.[206] What does the networked future that ANT envisions look like?

The relational ontology of ANT conceives the world as an assembly of associations, replacing traditional substance ontology, which conceived of the world as a totality of things. For ANT, meaning is not something that is added on to things by cognitive acts, signs, mental states, or language, but that which things are. The world of meaning is a universal process of networking and information construction in which associations are constantly being constructed, transformed, scaled up or down, extended, connected, etc. Networks are not things, even ordered compositions of things, as systems theory sees the world, and they are

[205] The status of science itself is an important question in both systems theory and ANT. Luhmann's sociology hardly fits into the functional subsystem of science whose true/false coding demands methods of verification. And Latour's vision of an associology raises the question of the status of the associations it itself is making, whether they belong to what Latour (2013) calls the specific "mode of existence" of science (REF) or to some other mode of existence, for example that of networking (NET), or are generally applicable to all modes of existence thus forming a metalanguage.

[206] This is the explicit program of Latour's *An Inquiry into Modes of Existence* (2013).

not cultural or social things as opposed to natural things. Networks are processes of networking. Just as life may be seen as a process of self-organizing autopoiesis, meaning is a process of networking. The network is the actor *in action*. ANT offers an alternative to traditional ontologies and general systems theory by proposing a theory of network order according to which networks are not closed, functional entities but open, flexible, scalable, and dynamic forms of order that can be defined as information. This takes up, extends, and grounds Luhmann's notion of society as a system of meaning but interprets communication, which for Luhmann is the operation of the social system, as processes of information construction through networking. ANT does not claim that Luhmann's description of autopoietic, informationally and operationally closed systems is false. Systems theory can usefully model certain forms of functional networks within society. ANT refers to such structured networks as "black boxes," that is, fixed input/output systems.[207] But what is true of the functional subsystems need not be true, as we saw above, of society as a whole. If the whole of society cannot adequately be modeled as an automated input/output system. What is it then?

With the close empirical observation and description typical of ethnological fieldwork, Latour meticulously followed the scientists' day-to-day work in laboratories.[208] He soon discovered that the idea of science as a closed system in Luhmann's sense, a domain that is clearly distinguished from other social domains, did not stand up to scrutiny. What the ethnologist investigating laboratory life and science in the making sees is

> ... visits by a lawyer who has come to deal with patents, a pastor who has come to discuss ethical issues, a technician who has come to repair a new microscope, an elected official who has come to talk about voting on a subsidy, a 'business angel'

---

[207] See Latour (1999:304) "When a machine runs efficiently, when a matter of fact is settled, one need focus only on its inputs and outputs and not on its internal complexity." This could be said of Luhmann's functional subsystems. They are efficiently running communication machines.

[208] See Latour for example *Laboratory Life* (....), *Science in Action* (.....), but *also An Inquiry into Modes of Existence* (2013).

> who wants to discuss the launching of a new start-up, an industrialist concerned about perfecting a new fermenting agent, and so on. (Latour 2013:30)

To assume that all these people from different social subsystems do not *communicate* with each other but go to great trouble to come together only to "disturb" each other, as Luhmann's theory would have us believe, is, no matter how much "structural coupling" we might assume, hardly credible. The scientists themselves insist that all these different people with their very different goals and concerns are necessary to the laboratory's success, and they are not at all interested in drawing boundaries or excluding everything that is supposedly not "science." On the contrary, the actual practice of science actively seeks and maintains associations with many other different social activities. For a laboratory to function, contrary to Luhmann's banning of all non-scientific communication into the environment of the science system, the laboratory depends on communication and close cooperation with many others. For example, the operation of laboratories in universities is closely linked to educational research programs, which are not directly concerned with constructing scientific proofs but with certifying skills, with business investors concerned with knowledge transfer, innovation, and profits, or with regulators involved with policies, and civil society actors motivated by public interests. Laboratories in the private sector operate in close cooperation with business, market pressures, competition, government regulation, civil society watchdogs, etc. Finally, religion and politics play a role in what science investigates even down to methods and theories. Science, in short, is not just science; it is business, education, politics, religion, law, and even art, but of course, doing experiments in a laboratory is not the same as contributing to a dossier on proposed regulation or designing a curriculum for a masters degree in engineering. These are all different forms of constructing different kinds of information. Nonetheless, the realities of science in action imply that Luhmann's view of society as consisting of autonomous functional systems informationally closed to one another does not tell the whole story. In practice, science and society resemble much more closely an open network with many crossovers and associations than a collection of autonomous, operationally and informationally closed systems that are more or less successfully adapted or structurally coupled to each other.

The ethnologist also soon discovers that scientific experimentation cannot adequately be described as verification or falsification of hypotheses about the objects under investigation. Bacteria or molecules are not passive objects deterministically reacting to the activities of researchers. The very meaning of the highly valued "objectivity" sought in science consists of letting the objects of inquiry also have something to "say" about the conditions under which they manifest themselves and the roles they play.[209] The actual work of experimentation is more like a set of challenges in which the objects of study are artificially enabled to act in specific, and often unexpected, ways by going through many "mediators," such as microscopes, particle colliders, or measurement devices, and various kinds of inscription and documentation. The results of an experiment are never immediately apparent, and no matter how often they are repeated, the results and interpretation could always be challenged and changed. The much-celebrated characteristic of "objectivity," what for Luhmann is the effect of the specific true/false coding of the science system, itself implies the object does not deterministically do what is expected of it, or what it is "told" to do. The object must be brought to say something about itself through the mediation of many artificial instruments and settings. Experimentation is, therefore, rather like a process of negotiation by which roles are discovered and assigned among human and nonhuman actors, negotiations of a specific kind in which both sides set conditions, show activities, demonstrate appearances and capabilities, and "mediate" the construction of information in many different ways. For this reason, Latour suggests that the traditional self-understanding of science as active subjects capable of a pure knowing of passive objects or so-called "facts" does not do justice to the empirical practice of scientific work, which is "coded" by variegated practices of inscription, documentation, demonstration, experimentation, etc. which Latour calls chains of "reference" or chains

[209] Latour has demonstrated this clearly in his work on Pasteur, The Pasteurization of France (1993).

of "immutable mobiles."[210]

Far from being a closed system processing only communications that can be either true or false, science in action looks much more like an open network of heterogeneous and hybrid actors pursuing different and often conflicting goals. The laboratory is not defined as scientific by means of excluding everything legal, educational, economic, political, religious, artistic, etc., but by including actors that are scientific AND legal AND economic AND educational AND religious, etc., but in its unique way. The actors involved are not merely human but also nonhuman, not only cognitive or communicative but also material, technical, and institutional. The network does not operate self-referentially like a closed system but expands and retracts in all directions simultaneously, depending on the changing needs, purposes, and programs of action of all the actors involved. In distinction from systems, networks are principally open in all directions and tend towards increasing rather than reducing complexity.

This brief review of how both Luhmann and Latour distance themselves from modern sociology raises the question of the status of the social sciences, at least, the status of these new theories. Are these theories considered "science" in the sense of that particular social activity concerned with generating "knowledge?" Are we still dealing with "science" as it has been traditionally understood in Western modernity? Or do the theories that Luhmann and Latour offer mark a paradigm change in the social sciences such that one must ask anew what science is and what role it plays in society? Indeed, in what kind of society does science play the role it does? The critique of the modern philosophy of science and epistemology, as well as the doubts both authors express towards modern sociology, seem to indicate that the theory of social systems and ANT demand a reappraisal of the status, function, and nature of scientific knowledge, or at least the status of their metatheories. An answer to this question rests upon the answers these theories give to how society is differentiated into various functional subsystems or different forms of networks.

[210] See Latour (2013).

We have seen that, for Luhmann, science is a functional subsystem delimited by its unique binary code of truth/falsity. The system of science consists of only those communications that can be coded, with appropriate methods, institutions, organizations, etc., as either true or false. All other communications, such as those dealing with whether something is legal or illegal, bought or sold, relevant for the certification of skills or not, etc., are banned into the environment of the functional subsystem of science. They perturb science, but they are not scientific. Despite the fact that sociology is only one discipline within the functional subsystem of science, Luhmann reserves a special place for what he calls "supertheories,"[211] that is, those foundational theories that serve as paradigms in the Kuhnian sense and are not subject to falsification. Supertheories are neither true nor false. They set the conditions or parameters within which anything can count as science. They function as the code, the borders of the science system.

As a supertheory, Luhmann's theory of social systems is, therefore, not strictly speaking science. It does not fall clearly within the boundaries of the system of science. Instead, it delimits and differentiates social communications that can count as science. The differentiation of communication into functional subsystems, one of which is science, is an autocatalytic, self-organizing, and constructive process within the social system. It is the self-differentiation of the meaning system. The supertheory cannot be understood as transcendental knowledge of the conditions of the possibility of science, such as the self-reflection of a transcendental subject, as modern epistemology since Kant has dealt with this issue.[212] Science claims universality since there is nothing that cannot be scientifically investigated, not even science itself. However, it achieves universality only by differentiating itself from other functional

---

211 Luhmann (1995:4-5) "Supertheories are theories with claims to universality (that is, to including both themselves and their opponents). If a supertheory achieves a significant centralization of difference, then a paradigm change also becomes possible. Systems theory is a particularly impressive supertheory."

212 Luhmann (1992:13) rejects modern subjectivism and thus also the transcendental/empirical distinction. The meaning system attempts to make itself meaningful as a system. We have argued that is questionable, since meaning may not be a systemic form of order.

subsystems and society as a whole. The same, of course, could be said of all the functional subsystems. Luhmann can claim, and he does, that the theory of social systems does what all scientific theories, at least foundational theories, do: it offers a framework for heuristics, hypothesis construction, and research programs. While this is, to a limited extent, valid, the universal claims and the level of abstraction of the theory point toward a paradigm going beyond the boundaries of normal science. If the theory of social systems is not clearly located within the system of science, where is it, and what function does it fulfill? Luhmann locates his theory on the level of a supertheory that itself is neither within nor without the system of scientific knowledge it supposedly explains. We have argued above that this mysterious form of knowing, in all its forms, cannot explain itself as long as it assumes that it is a system. We have argued that with Luhmann, meaning, as it were, misunderstands itself as a form of systemic order. In short, we claim that if meaning knew what meaning is, it would understand that it is not a system. A theory of meaning is meaning's self-explication. Is a self-understanding of meaning science? For general systems theory, this appears to remain an open question. How do things stand with ANT?

At first, it seems that ANT is only concerned with how everything is connected to everything in networks; that is, ANT seems to emphasize the continuity of the social and not its differentiation. Latour is openly critical of talk of "domains" that would break up the collective by distinguishing society from nature, subjects from objects, knowledge from opinion, micro form macro actors, structure from agency, and the transcendental from the empirical.[213] ANT emphasizes the continuity between networks to clearly outline a theory of society as the universal collective of humans and nonhumans. Society, or the collective, indeed, the world, is through and through networking. There is nothing that is not networks networking. How, then, does ANT answer the question of social differentiation? Is the collective differentiated at all? And if so, how does this influence the status of ANT and the social sciences?

[213] Interestingly, Latour seems not directly to have commented on Luhmann's theory of functional subsystems or taken up a dialogue with systems theory.

3.6 Network Differentiation

Latour's description of society from the actor-network perspective is presented in his major work, *An Inquiry into Modes of Existence* (AIME) (2013).[214] The work is subtitled *An Anthropology of the Moderns,* which indicates an empirical and descriptive methodology and emphasizes that the goal is not to offer a general theory of society but a description of modern Western society. Latour's stated purpose is to describe the "Moderns" much as an ethnologist doing fieldwork with a tribe in Borneo would do, that is, to empirically describe the Moderns such that what they truly value comes to the fore, apart from their often misleading beliefs about themselves. An ethnology of the Moderns is needed because it allows the Moderns to see themselves in a different light, from the perspective of an "objective" description of their practices and beliefs.[215] This is important for Latour because it supposedly allows the Moderns to enter into a fruitful and promising dialog with other cultures, a dialogue which the misleading and arrogant self-understanding of the Moderns has hindered for centuries. The unprecedented problems facing society today are global: climate change, migration, pandemics, financial insecurity, etc. The age of global dialogue on common issues, above all, the climate crisis, has come, and modern Western society is no longer in a position to dictate the terms upon which a global solution could emerge. The description of modern society that Latour develops under the title of *An Inquiry into Modes of Existence* (AIME) seems, therefore, not to have a purely "scientific" motivation. It is not merely a question of whether the description of the Moderns he offers corresponds to how the Moderns are, but it has a normative purpose. It aims to enable cultural comparison, mutual understanding, and global negotiation such that the problems facing humanity today and in the future can be addressed cooperatively and thus effectively. In addition to this explicitly normative purpose, the main title of the work suggests that it is concerned with philosoph-

---

[214] This is the main publication of an EU funded research project whose Website is http://modesofexistence.org/.

[215] "Objective" in this context means that the self-description of the Moderns is not taken at face value.

ical questions and not merely empirical descriptions. An inquiry into "modes of existence" suggests ontological and metaphysical concerns, usually considered philosophical and not scientific. Therefore, the reader is confronted right at the outset with whether Latour's enterprise is science or philosophy or something else altogether.

Latour quite openly admits that his enterprise is not normal science and not even the attempt to establish a paradigm as Luhmann understands his project of a "supertheory." Latour himself expresses uncertainty about the status of his theory. While he admits that his inquiry into the beliefs and practices of the Moderns is not normal science, he has no clear answer about what it is, even though the role it is intended to play in society seems to be clear.

> ...it is not in the mode of knowledge that I claim to be working. The term "inquiry" has to be taken in a plurimodal sense whose object is to preserve the diversity of modes. Can we call this approach "empirical philosophy"? I am not sure, given how indifferent philosophy has become to the tasks of description. Experimental metaphysics? Cosmopolitics? Comparative anthropology? Practical ontology? (Latour 2013:481)

Whatever one calls it, it seems clear that for Latour, science in general and his own theory is an endeavor that serves practical purposes. It has a goal beyond objective knowledge of how modern society is differentiated into what Latour calls "modes of existence." Of course, because it is "construction," all information construction has a built-in value judgment. The idea that information must be "constructed" in the threefold sense of distributed agency, uncertainty of outcome, and valuation mentioned above prohibits any assumptions about science being a pure and disinterested knowing of objective facts. At the same time, it implies that all the modes of information construction are goal-directed in that they are "obliged" to construct well and not badly. Science in action, for example, is a process of translating and enrolling actors in actor-networks in ways directed to constructing objective knowledge. This goal is what Latour calls a "value," and practices of networking that are held together by being directed toward a particular value show that we are dealing with a specific "mode of existence." As a mode of existence, science, for example, is not to be confused with other networking practices such as law, religion, business, politics, etc. These different practices of networking are directed toward their own values. If we ask where in this framework of modes of existence we can find the kind of science that Latour himself is doing, we must ask what its goal and value are. The value that Latour strives for is global cooperation based on mutual understanding. Is this science? If not, what is it? And what is the connection between the objective knowledge about the Moderns revealed by the ethnological investigation and the purpose of global negotiation guiding the inquiry? One must ask of Latour the same question he asks of the Moderns. How does the self-description of ANT compare to the description ANT offers of science? Is ANT itself not a form of networking? If so, how is it different from science? If it is different from science, what is it, for it is certainly not religion, politics, business, or any other of the various modes of existence?

For Latour, science consists of a particular kind of networking that mobilizes artifacts or "inscriptions" in relation to each other such that the content transported by these artifacts remains the same throughout long chains of reference. The artifacts can be anything: a diagram, a measurement, a graph, a sample, a demonstration, a textual document, images, models, simulations, or anything that can move information without changing it. Fittingly, Latour calls such artifacts or inscriptions "immutable mobiles" and locates their purpose in making remote objects accessible. For example, a carefully arranged set of earth samples in a wooden lattice that is assembled in the field "represents" the exact relations and content of that area of land under investigation when transported hundreds of miles away into a laboratory. The assemblage, the artifact, is mobile, but it transports the original object, the relative composition of the soil, unchanged or immutably to a distant place. In the laboratory, the soil samples are chemically analyzed utilizing special instruments whose results are transported to a graph, which is then compared to other graphs to end up in a scientific paper published in *Nature* or on a PowerPoint slide presented to students, government agencies, or business developers. Throughout this entire chain of references, the remote object, that is, the composition of the original soil, must be kept constant. One must be able to follow the chain of references backward to the original land and "reproduce" the results. Otherwise, what comes out at the end is not "objective." Only by means of such chains of references among immutable mobiles can "objective knowledge" be constructed. Assembling long chains of such immutable mobiles is the specific mode of networking (Latour speaks of "mode of existence") of science as distinct from law, religion, politics, art, etc., which all have their own ways of constructing information. But is this the science that Latour himself is doing?

On the one hand, Latour claims to offer careful empirical descriptions of the experiences the Moderns make in their various modes of existence. These phenomenological descriptions can be "verified" by all readers. Every reader has the right and duty to ask: Does this description fit with my experience? Science and Technology Studies has operated successfully with this methodology for decades, and Latour is convinced it has passed the test of scientific verifiability. Luhmann would probably agree and say that science is a functional subsystem operating according to its own code of true/false, distinguishing scientific communication from other kinds of communication in other social subsystems. But for Latour, science is not an operationally and informationally closed system. It is an actor-network and is open to information from all other networks. The collective is not ordered in the same way that Luhmann describes in terms of functional differentiation. There is more to the collective than the functional "black boxes" that are autonomous social subsystems. In order to express this idea, Latour (2004a; 2005:98ff.) has introduced the term "matters of concern" in opposition to "matter of fact."

Matters of concern are not the objective facts of modern science, where the term "objective" means that nothing is coming from a subject, and nothing is artificial, social, or *constructed* about them. The objective facts of modern science exist simply as they are in the ontological domain of nature until science somehow, without any mediation, construction, or social effort, "discovers" them. Matters of concern, on the contrary, are those facts (matters) that only reveal themselves by mediation and construction. Without "concern," without the "gathering" of many different actors, trajectories, contingencies, and uncertainties constantly being disputed and revised, and without the unpredictable crossings into different networks, no scientific fact could come into being. To be precise, the objective knowledge Latour identifies as the "value" or goal of the mode of existence called science is neither natural nor social, neither subjective nor objective, neither artificial nor natural. These typically modern distinctions are what the ethnological description of how science works places into question. According to Latour, they are to be replaced by the question of whether knowledge has been constructed well or badly, whether all relevant actors have been taken account of and "gathered" together properly.

What does this mean for the status of science and its role in society? As a practical activity, science, like all other practical activities, can be considered from the perspective of "design."[216] Science, for Latour, is not confined to an operationally and informationally closed subsystem of society concerned exclusively with distinguishing between truth and falsity. Recall the above-cited example of the laboratory director who must deal with law, business, government regulations, educational curricula, private donors, logistic chains, machines of all kinds, and much more to do science. The anthropologist of the Moderns sees that science is intimately bound up with many other social activities and with many nonhumans. Science is not pure knowing, somehow performed by a disinterested observer, which is how the Moderns mistakenly think of science. It is practically involved in constructing the many programs of action that influence the conflicting trajectories of the global network of humans and nonhumans, or the "collective," as Latour calls society. For Latour, the scientist, the matter of concern that

216 The knowledge discovered by science helps in the "design" of institutions. "... the proposition I am exploring through this inquiry consists in using a series of contrasts to distinguish the values that people are seeking to defend from the account that has been given of them throughout history, so as to attempt to establish these values, or better yet to install them, in institutions that might finally be designed for them" (Latour 2013:7). For a reinterpretation of action in general as "design" following Latour see Belliger/Krieger (2021).

must be appropriately gathered is the collective as a whole. When it comes to social differentiation, the question is not to describe functional subsystems as Luhmann does, but to gather the collective in the right way, to make the collective the most important "matter of concern." There is a direct connection between striving for objective knowledge (the mode of existence of science), which distinguishes the various modes of existence, and concern for the collective, which strives to gather all things together.[217]

Corresponding to these two directions, it is not surprising that Latour explicitly acknowledges the philosophical character of his project despite the emphasis on empirical description. There is a tension between title and subtitle because the title speaks of modes of existence and awakens expectations of a philosophical ontology or metaphysics, not merely empirical anthropology. Indeed, Latour claims to offer a "pluralistic ontology." This claim is directed against a certain kind of metaphysics, which has always been interpreted as a general theory of Being applicable to all beings no matter how they differ. According to this view, reality is one, but there are many ways of speaking of it. Latour, the empiricist, has never been comfortable with universalistic claims of philosophy and especially metaphysics, which, as he puts it in a well-known self-criticism (Latour 2013:35), ends up "saying the same thing about everything." If everything is a network, how are things different? There are many concepts in ANT besides the concept of "network" that seem to apply to all entities, for example, "translation," "enrollment," "mediation," and "agency." In our reading of ANT as a theory of meaning, we propose not only adding a term Latour does not use but also openly admitting that this concept is metaphysical. This is the concept of "information." We use this concept to describe what networks consist of, that is, relations/associations as defined by Latour's principle of irreduction. We interpret the principle of irreduction, admittedly against Latour's will, as a metaphysical statement saying the same thing about everything. What it says about everything is that it is a relation, which we propose to call information or meaning.

[217] One recalls the scholastic principle "distinguish in order to unite."

This position allows us to interpret Latour's modes of existence *as modes of information construction* or *modes of networking*. Therefore, the philosophical program of ANT is a contradiction since it proposes at once to say the same thing about everything while insisting that being is plural. It would seem that, according to Latour, there are different modes of existence, but no one Being.[218] How can this contradiction be resolved?

Let us turn first to the differentiation of the collective and then return to the problem of unity. For ANT, society is differentiated not by functional subsystems but by different networks of information construction. They are all networks and share the constitutive principles of networking, but they are different networks insofar as they each have their own "mode of existence." As networks and not systems, they are all operationally and informationally open to each other. In contrast to systems, networks pursue multiple goals and diverse programs of action. In contrast to systems, networks are constantly changing, scaling up or down, operating at both global and local levels. Instead of a society differentiated into autonomous functional subsystems, which Luhmann describes, we have a "collective" of humans and nonhumans that is differentiated into different forms of networking and different modes of information processing. Since information is Being, these networks can be called "modes of existence." Where Luhmann speaks of functional subsystems, Latour speaks of modes of existence. Like Luhmann, for whom there is only one world society, for Latour, the collective is global. But against Luhmann, this "post-modern" society is characterized by "network differentiation" instead of functional differentiation. Luhmann often seems not to fully realize how far the social systems theory has moved beyond modernity, for he characterizes "modern" society by functional differentiation. For Latour, "post-modern" society is defined by network differentiation. What is network differentiation from the point of view of Latour's description of modes of existence?

There are many points of comparison between Luhmann's functional

---

[218] We do not attempt to resolve this paradox but by means of our interpretation of ANT as a theory of meaning and by emphasizing information we propose to simply side step it.

subsystems and Latour's modes of existence. The vocabulary is different, but often something quite similar is intended. Instead of functions, Latour speaks of values. For example, the "value" embedded in scientific practice is "objective knowledge."[219] The ways in which objective knowledge is obtained are called the "conditions of veridiction," the specific ways scientific truth is constructed by means of chains of reference. These conditions of veridiction are different for law, religion, or politics, just as the programs of action, values, or goals these networks pursue are different. For religion – Latour speaks only of Christianity – the mode of veridiction is the experience of being personally addressed by a saving power. For law, establishing a legal judgment depends on marshaling the necessary "means." For politics, it is the continuously renewed assembly of all participants negotiating an issue. For art, what Latour calls "fiction," there is the appearance of form in a material of some kind. For organizations, there is the narrative ordering of actors, roles, and purposes by means of at once authoring and being subject to "scripts." For individuals, there are the many ways of dealing with those transformative forces guiding or hindering psychological development. For technology, there are many devices, detours, and delegations of action onto artifacts of all kinds. For social structures, there is what Latour refers to as "habit," the automatisms of behavior, belief, and action that take over as soon as one does anything on a regular or continuous basis and conforms to expectations. For business or the economy, Latour speaks of "attachment" or "passionate desires" for things. Interestingly, he does not mention money or anything like Luhmann's symbolically generalized media. For morality, Latour identifies "scruples" about whether any information construction has been done well or badly. Since construction necessarily involves the value question of whether something has been constructed well or badly, morality arises as a mode of existence from the scruples one must have about all actions.

In all of these modes of existence, we have no difficulty recognizing topics that have long preoccupied sociology, psychology, economics,

[219] Latour reserves the concept of "truth" to describe the "felicity conditions," or the way in which each mode of existence retains its unique practices in the achievement of its goals or it value. All modes of existence have their own "truth."

political science, ethics, and other human sciences in one way or another. What strikes the reader as unusual in Latour's description of these well-known phenomena is that although he focuses on how we subjectively *experience* them, the phenomena themselves are granted a peculiar ontological status as distinct modes of existence. There is a strange identity between experience and being; for example, moral scruples are not merely subjective experiences people "have" but beings that have a grip on people. Forces of psychological development and transformation are not internal, mental experiences that people "have" but beings that hold sway over people. The passionate interests humans have for things of all kinds that drive production and consumption in the economy are not merely strong emotions but beings in their own right. This surprising and disconcerting confusion of subjectivity and objectivity is, of course, intentional. It aims to overcome the Modern's false understanding of themselves as subjects distinct from objects, agents separate from a passive world of things, and belonging to a social and cultural realm distinct from material nature.

The investigation into modes of existence ends with a list of fifteen modes or different kinds of beings.[220] The fifteen modes, a list which Latour admits is provisional, are divided into four groups. Latour distinguishes between modes of existence that, in his view, are more related to "quasi-objects," such as Technology, Fiction (Art), and Reference (Science), and those related to "quasi-subjects," such as Politics, Law, and Religion. In this context, the term "quasi," which Latour takes over from Serres, refers to the assertion that these beings are neither subjects nor objects, neither purely social nor purely natural, but somehow both together. Of course, this can be said of all the modes of existence. Nevertheless, there are also modes of existence that are unrelated to subjects or objects. These are Reproduction (i.e., the subsistence of material things and viability of living organisms), Metamorphosis (i.e., forces of psychic development and transformation), and Habit (i.e., behavioral automatisms, social structures, frames). There is a fourth group of beings that link quasi-subjects and quasi-objects.

---

[220] See Latour (2013:488) for a table listing all the modes and their respective characteristics.

These are Attachment (i.e., business, the economy), Organization (i.e., the narrative ordering of roles, times, and places), and Morality (i.e., scruples about whether construction is well done or not). And finally, there are modes of existence that apply to all others and, thus, to all beings. These are Network (i.e., entering into associations), Preposition (i.e., marking the distinctions between modes), and Double Click (i.e., that mode which covers over the networking to which all beings owe their existence as well disguising thereby the distinctions between modes). Double Click, a reference to a computer mouse's commands that initiate and conceal complex processes, is responsible for the typically modern self-misunderstandings of Subject/Object, Nature/Society, Agency/Structure, Individual/Society, Fact/Value, etc.

Latour's ethnology of the Moderns and their practices has revealed more and different modes of existence than the functional subsystems Luhmann describes. For Luhmann, modern society consists of the functional subsystems of science, law, politics, religion, business, education, media, art, and healthcare. Luhmann also considers organizations, which are found within all subsystems, to be systems in their own right, and finally, even face-to-face interactions are considered a form of systemic order. Latour remarkably omits healthcare, education, and media but characteristically adds the way of being of material things, such as stones, mountains, rivers, etc., and of living organisms, which Luhmann explicitly excludes from the social system. Luhmann also excludes technology or any artifact from social participation. After all, at least until recently, tools and machines do not communicate. For Latour, all tools and machines, indeed all artifacts such as the technologies of language, writing, imaging, and so on, are social partners contributing in their own ways to networking, that is, to the construction of information. Networking itself is dealt with as a separate mode of existence, but it is a special way of being since it is the basis of all modes. The collective as a whole can only be "gathered together," which is the very purpose of conducting the inquiry into modes of existence through networking. We need not go into each of these modes of existence in detail. As we noted, meaning can understand itself and differentiate itself as it wishes. Our concern is whether meaning understands itself better as a system or as a network.

It will come as no surprise that everything hinges on what is meant by the term "network." Let us cite a long passage, which we have partially quoted above that sets the stage for understanding this all-important term.

> Instead of wondering, for example, if Science is a domain distinct from Politics or The Economy or Religion, the investigator will be content to start with some arbitrary sequence of practices. For example, she goes into a laboratory: there she finds white lab coats, glass test tubes, microbe cultures, articles with footnotes—everything indicates that she is really "in Science." But then, with a certain obstinacy, she begins to note the origins of the successive ingredients that her informants need in order to carry out their work. Proceeding this way, she very quickly reconstitutes a list of ingredients characterized by the fact (in contradiction with the notion of domain) that they contain ever more heterogeneous elements. In a single day, she may have noted visits by a lawyer who has come to deal with patents, a pastor who has come to discuss ethical issues, a technician who has come to repair a new microscope, an elected official who has come to talk about voting on a subsidy, a "business angel" who wants to discuss the launching of a new start-up, an industrialist concerned about perfecting a new fermenting agent, and so on. Since her informants assure her that all these actors are necessary for the success of the laboratory, instead of seeking to identify domain boundaries, which are constantly challenged by innumerable erasures, nothing prevents her any longer from following the connections of a given element, it hardly matters which one, and finding out where it leads. (Latour 2013:30)

From the network perspective, science does not consist of the scientific method and its epistemological assumptions or any theory of science as the truth about facts that speak for themselves. Besides the careful inscription of experimental results and measurements, science consists of many other things: artifacts, machines, institutions, regulations, supply chains, funding procedures and conditions, social concerns, business applications, etc. All these things and artifacts are subsumed under the mode of existence of Technology. Interestingly, any of these "things" can become the starting point of a network of associations that branches off into all directions into society; indeed, if one goes far enough, they come to participate in all other modes. If one follows the trail long enough, anything can be associated with almost anything else. Let's go back to our assumption that this is all information construction, and it is the nature of information to be able to be linked with all other information. Networking is what the collective in all its forms consists of. For Luhmann, this situation creates the problem of over-complexity of communication that must be reduced or channeled in specific ways. Luhmann solves this problem by introducing the idea of symbolically generalized media and functional subsystems. For Latour, there arises a similar problem of describing how the global network of all things and practices is reduced to forms recognizable as particular social practices such as science, law, politics, or religion. How is the unity of the global network, the "world," differentiated into specific forms of networks? Latour makes the theoretical move that makes this question possible at the outset by distinguishing two meanings of the term network.

On the one hand, there is the network of things and practices that are all associated in ever-widening circles that the anthropologist describes. This network is global and cannot be limited to any "domain" of society, whether law, politics, economy, religion, etc. But on the other hand, there is the network of what flows through any particular setup of things and practices. Focusing on what is flowing rather than the fact that it is flowing allows Latour to make an important distinction. This distinction is borrowed from that which serves as the basis for his pluralistic ontology, namely, the distinction between how a network is set up and what flows through the setup of the network. Latour (2013:32) cites the example of a mobile telephone network:

> The fact that information can circulate by means of a cell-phone network tells us nothing about the way the network has been put together so as to work, right now, without a hitch: when all

> the elements are in place and everything is working well, in the digital window of our cell phones what we can track is only the quality of a signal marked by a certain number of rising vertical bars (by convention, from one to five). The "network" in the usual sense of technological network is thus the belated result of the "network" in the sense that interests our investigator.

And he goes on to add:

> The distinction between the two senses of the word "network" would be the same if she were interested in railroads: following the tracks is not the same as investigating the French national railroad company. And it would still be the same if, taking the word more metaphorically, she wanted to investigate "networks of influence": here, too, what circulates when everything is in place cannot be confused with the setups that make circulation possible. (Ibid)[221]

And he concludes:

> So under the word "network" we must be careful not to confuse what circulates once everything is in place with the setups involving the heterogeneous set of elements that allow circulation to occur. (Ibid)

What the Moderns have done is systematically ignore (Double Click) the "setup" and create theories explaining what flows, whether it be the flow of truth, of legal decisions, of contact with transcendence in religion, of collectively binding decisions in politics, of originality in art, or of business in the economy. The setup, the network of heterogeneous associations, the myriad of things, artifacts, institutions, organizations, etc., all linked together such that a particular kind of flow is possible (Latour speaks of "pass" through the gaps between the heterogeneous

---

[221] It should be noted that for network science, for example, in the investigation of internet email networks, the nodes, whether persons or servers, as uniformly defined at the outset and not further tracked into all the many other associations and components make constitute them and same is for the links.

entities, or "continuity" preserved through discontinuity), has always been there as the condition of the possibility of what flows.[222] Throughout modernity, the first meaning of the term network, the setup, has been enhanced, expanded, and deployed globally. What the Moderns have actually done with their science, technology, and business is to increase the number of hybrids enormously. The networks of humans and nonhumans have dramatically expanded. This is the setup. It is through these networks that everything, money, goods, knowledge, people, etc., flows. But the account of the flow, the second meaning of the term network, has skipped over the setup and gone directly to the results of the flow. To a certain extent, the setup is itself responsible for its own forgetfulness. It is because the network of associations of heterogeneous things has been painstakingly and laboriously set up so well and efficiently that it disappears, and only the results are visible. One sees, for example, that there is knowledge of objective facts, but one has lost sight of all the detours via things, instruments, inscriptions of all kinds, and the institutions necessary to arrive at this knowledge. If one leaves all the detours of the setup out of account, there appears only a subject knowing an object, and there appears a material world of nature as opposed to a world of signs, meaning, culture, and society. Out of this forgetting of networking, a distinctively "modern" worldview has been created with clear distinctions between society and nature, subject and object, mind and matter, language and reality, individual and society, freedom and determinism, rationality and belief, etc. Once the project of the anthropology of the Moderns has been successful, and the dependence of the flow upon the setup has come into view, the Moderns should recognize, as Latour provocatively puts it, that "we have never been modern."[223]

---

[222] See Latour's (2013:33) definition of the term network: "The notion of network can now be made a little more specific: it designates a series of associations revealed thanks to a trial—consisting in the surprises of the ethnographic investigation—that makes it possible to understand through what series of small discontinuities it is appropriate to pass in order to obtain a certain continuity of action."

[223] We are not concerned in this book with Latour's "archeology" of modern ideas, but with the network paradigm or order that emerges from ANT and will therefore not rehearse the archeology of modernity that takes up most of the AIME project.

What distinguishes ANT from systems theory concerning social differentiation is the fact that the various networks, because they are all interconnected, "cross" each other, creating thereby a tapestry or tissue of trajectories of information construction that pursue multiple goals, are open on all sides and in all directions, expand and contract, and are constantly changing. This network of networks is society, or, following Latour, the collective. To be "collected" differs from being selected, related, controlled, and excluding an environment. Being collected is different from being selected, whether as an element of a system or a system under the selective pressure of the environment. Networks do not arise in an ever more complex environment to which they must adapt by viably reducing complexity. Comparing Luhmann's theory of social systems and ANT, we see that in the place of a view of society as consisting of autonomous functional subsystems that are merely structurally coupled or adapted to each other, there appears the collective of crossing and linking networks. It is these crossing networks that constitute society as a whole. A global network society, however, is not bounded or contained within any system boundaries. ANT has no problem with an unbounded collective. However, the theory of social systems runs into conceptual difficulties when society must be a bounded system.

For Luhmann's theory of social systems, there are no processes of selection, relationing, and steering that order society as a whole, for there can be no system of everything. On the other hand, Latour's technical mediation and processes of translation and enrollment are ubiquitous. They must include everything, or at least they should. Other forces and norms weave the social fabric together rather than merely functional dependence, understood as the need for an autonomous system to adapt to its environment. For Luhmann's functional subsystems, other systems are only environment, to which any system must be structurally coupled to continue its autopoiesis successfully. This is what it means for a social system to be "functional." A network must not be functional. It can be. Functional networks are referred to in ANT as a "black box." Black boxes can be opened and transformed when actors become mediators again, and information construction begins anew.

Functionalism implies that a social system can become dysfunctional, as, for example, is increasingly the case for the political system, which is supposed to provide collectively binding decisions but can do this only within a nation-state, which is no longer what the "collective" is. When the collective has become global but organized on the level of territorially bounded nation-states, politics can no longer fulfill its function of steering society. Instead, we are witnessing a Hobbesian a-social state of nature and a war or all against all on the international level. It is increasingly apparent that the political system cannot effectively address any present-day global problems. The climate crisis is a case in point. On the contrary, the networks described by ANT are unbounded, scalable, and continuously crossing each other, participating in each other, thus making up the social understood and experienced as the "world." For this reason, Latour is convinced that the network view of society can lead to a new, or "ecological," form of global cooperation. Politics, for example, can become "political ecology."[224] Global governance need not be a utopian dream.

The network view of society comes to the fore in the example cited above of the laboratory, where, as we saw, not only science but law, education, business, etc., come together to make the laboratory a successful scientific institution. The laboratory illustrates what this change of perspective from systems to networks means. Even though the description focuses on science, there is communication and cooperation with lawyers, business people, regulators, religious believers, students, technologists, etc. This complex, multi-network communication is constantly contributing to and shaping the research that is going on in the laboratory. Furthermore, there is an ongoing interaction with nonhumans of all kinds whose contributions are necessary for the construction of information. What is going on at the laboratory cannot be adequately modeled as an operationally and informationally closed communication system being perturbed and disturbed by meaningless interventions from various environments, forcing the scientists to construct the information necessary to adapt to these environments

---

[224] See Latour (2004b) for an exposition of his political theory.

internally. Latour's detailed description of what actually happens in society is much closer to reality than Luhmann's view through the lens of the abstract theory of systemic order and the exclusion of nonhumans. The collective as a whole with all its different networks comes together in the laboratory, but of course, in a special way, different from how it comes together on Wall Street, or in Congress, or church, or at an art museum, or in the editorial office of a media company, etc.

The laboratory is indeed society, but society is "articulated" in a particular way such that the network setup has its own character, way of translating and enrolling actors, and methods of deciding what flows through the setup, that is, its specific value. Latour seems to be more concerned with the diversity of networks than their unity. The *Inquiry into Modes of Existence* ends with a list of fifteen different kinds of beings and not a unified vision of a global society. Of the two principal ways of ordering information, the list or the narrative, Latour concludes his investigation with a list. But we must not forget that this list appears within a broader narrative, the story of "gathering together" the global collective with the purpose of addressing the climate problem. In the end, despite the list, we are offered a unifying narrative. In this story, the fifteen different kinds of beings the investigation discovers take on roles and pursue goals beyond their individual networking forms. This is the story of a global network society. It is the story of a special kind of science dedicated to gathering together the collective. The social sciences, indeed, all forms of knowing, are taken to task for what and how they construct information. The supposedly value-free science of the Moderns is confronted with an ethical imperative to construct well. In this new narrative, the role of science in society changes. But it is not only science that is changing. New roles and new narratives are appearing throughout society. This is what we will examine below.

Part III

From Systems to Networks – A Conclusion and a New Beginning

Chapter 4

From Systems to Networks – A Conclusion and a New Beginning

Latour often speaks of the passing of Western modernity. He speaks of the closing of a parenthesis; of the end of a historical moment in which the Moderns not only realize they have never been modern but in which "society" becomes the "collective." We propose referring to this historical, cultural, and social event as the "digital transformation." Although this term normally designates the changes an organization goes through when adopting digital technologies, we use it in a much broader sense to designate the global, social, and cultural changes brought about by the technological advances of the 21st century. We are witnessing and participating in the emergence of a global network society. In saying that technology changes society, we are not returning to theories of technological determinism but referring to Latour's notion of technical mediation, which we have interpreted in this book as information construction. When our poor hominin, more than three million years ago, picked up a stone that became an ax and which transformed him or her into a hunter or a warrior, neither the hominin nor the stone was "determining" anything. There was no determination or causation from the stone or the hominin. Instead, there was a "gathering" of actors into networks and the construction of information or meaning. Instead of saying that technology changes society, we should follow Latour, who says that technology is society.

Luhmann describes a society as differentiated into functional subsystems. For Luhmann, this is a description of modern society precisely because functional differentiation succeeds stratified and hierarchical forms of social differentiations that characterize pre-modern societies. Latour describes a "collective" that is no longer modern Western society but a networked world, including humans and nonhumans, which, as we have seen, can no longer be adequately understood in modern terms. Latour attempts to describe this new, but, as we argued, very old, collective in terms of modes of existence. Latour's modes of existence are in some ways similar to Luhmann's functional subsystems. Both Luhmann and Latour are talking about a society in which science, business, politics, religion, education, etc. characterize social order and human endeavor. But just as Luhmann's functionally differentiated society, so too, Latour's pluralistic ontology of modes of existence does not tell us about society as a whole or what kind of society the collective is becoming. Luhmann leaves us with a society fragmented into autonomous, functional subsystems that do not communicate with each other but somehow exist in a state of mutual adaptation, and Latour leaves us with a plurality of modes of existence whose crossings and overlappings tend to cause more harm than good. Following our information theoretical interpretation of Latour, we propose to speak of a global network society. And we ask, what are this now emerging world's guiding concepts, norms, values, and practices?

The global network society can be described neither in terms of functional subsystems nor as consisting of different modes of existence, but instead, by reference to the guiding norms of networking. Networking, or the construction of meaning through translating and enrolling actors into networks, follows certain norms that guide how it is to be done well. Construction, as noted above, always involves a value judgment, the judgment of whether what has been constructed has been constructed well or badly. Latour argues that a scientific fact, just as much as a legal judgment, business deal, or work of art, can be constructed well or badly. This applies to all of Latour's modes of existence since all are forms of networking and information construction. We will argue that by focusing on the norms of good networking instead of the various modes of existence or functional subsystems of which society at any time might consist, we can understand society as a whole and attain a vision of where society is going. Our present historical moment calls for a vision of a global future. We assume this is what the entire "ecological" shift and the increasingly urgent call for global governance is all about. We, therefore, embark upon the admittedly dangerous and uncertain path. Still, it is a path that lies at the heart of Latour's notion of what social science today should be about, namely, to contribute to gathering together the global collective.

Instead of the functional subsystems that Luhmann assumes, or the modes of existence Latour describes, we propose focusing on what could be called "network norms." Speaking of network norms does not imply that society, or the collective, no longer consists of organizations, institutions, or systems such as education, business, science, healthcare, art, religion, etc. In Luhmann's view, what are necessary functions for social order and modes of existence for Latour are all still part of the global network society, but they are ordered differently. They arise from the constructive activities of networking that follow, more or less, the network norms we shall describe below.

We claim that the stone that became an ax in the hand of our ancient hominin appeared with the emergence of meaning. It appeared for the first time as something that "subsists" in its own way (what Latour calls the mode of Reproduction); it appears as something to which activities can be delegated (Mode: Technology); it appears as something whose repeated use becomes "second nature" (Mode: Habit), as something formed of material (Mode: Fiction/Art), as something that can carry knowledge (Mode: Reference/Science), as something about which all stakeholders dispute and negotiate (Mode: Politics), as something that testifies and bears witness to events (Mode: Law), as something to which many hominins (and humans as well) become passionately attached (Mode: Economy), as a bearer of scripts and narratives assigning roles to animals, enemies, wood, etc. (Mode: Organization), as a source of scruples about whether the ax has been constructed well or badly (Mode: Morality), as a ritual object mediating contact with transcendent beings (Mode: Religion), and as linked in many unforeseeable ways to more and more things as time goes by (Mode: Network). The emergence of the stone ax, which illustrates the emergence of meaning having many modes, is a "gathering" that happened way back then for the first time and has been going on ever since. This gathering is what it means to talk about the emergence of meaning.

Gathering is another word for constructing relations, that is, constructing information. Meaning is a level of emergent order that integrates both humans and nonhumans into actor-networks that exist as information, associations, or relations. Just as our hominin became a warrior or a hunter associated with the stone ax, we are becoming informational beings. Floridi (2014) speaks of "Inforgs" to designate humans in association with what could broadly be called digital technologies. Our lives no longer revolve around stone axes, which was the case for hundreds of thousands of years, but around computers. It is *technical mediation* that makes meaning; this is why technology, in the broadest sense, plays a major role in how the collective is gathered. The gathering is always technology, and technology is always a gathering. Discovering and describing the ways in which the affordances of digital technologies normatively influence and guide the activities of networking that are going on today is, therefore, the task at hand. It is the task that ANT has bequeathed us. It is what is left after sociology has ceased being one of the disciplines of modern science and become responsible for gathering the collective. It is what the paradigm shift from systems to networks is all about.

4.1 Digital Transformation

The digital transformation has disrupted Western modernity in at least three important ways. The first disruption is the posthumanist revision characteristic of both systems theory and ANT. The posthumanist revision no longer puts the autonomous rational subject at the center of history and society. The myth of humanism has become problematic in many ways, not the least of which is that it seems to have taken its last stand on the individual. In the digital age, individual freedom is often interpreted as an issue of privacy. Privacy is an attempt to block connectivity and the free flow of information characteristic of the digital transformation and its new norms and values. Privacy is how the bounded individual attempts to resist being transformed into information, that is, becoming a being that is no longer defined by individuality and freedom but by relationality and association.[225] Despite

[225] For a discussion of the meaning of privacy in ethics and regulation, see Belliger/Krieger (2018).

Luhmann's posthumanism, individuals are considered as systems with constitutive boundaries. For ANT, on the contrary, networks have no constitutive boundaries. Instead of privacy, the global network society values "publicy." Publicy is the default condition of the informational self, that is, the form of identity that has emerged from the affordances of information technologies. The informational self is essentially related, connected, and open. The informational self is not a bounded individual but a network of relations, that is, a being that exists embedded in actor-networks and, therefore, as information and not as substance.[226] The attempt to save the autonomous rational subject of humanism using privacy as a fundamental and inalienable right derived from an ahistorical human nature fails to account for history and, more importantly, the relational nature of information. Informational selves cannot be bounded individuals constituted, as a recent interpretation would have it, by information that can in no way be shared, communicated, or used in social interaction.[227] The informational self exists in the condition of publicy and not privacy; that is, it exists as connected to many other actors, both human and nonhuman, in many different actor-networks. The disappearance of the autonomous rational subject of humanism is the first disruption characterizing the digital transformation and the advent of a global network society. It is the disruption of human self-understanding, at least for Western modernity.

If the first disruption is the loss of the myth of humanism, which, as Latour might say, could be called the discovery that we have never been modern, the second is the loss of an age-old principle of social organization, namely, hierarchy. Since the earliest times, we have grown accustomed to organizing cooperative action among large groups by means of hierarchy. As soon as more than twenty or thirty people get together, communication's spatial and temporal constraints require that someone be the boss, the chief, the king, the leader, the president, etc. As Shirky (2008) has noted, face-to-face communication does not allow large groups to coordinate group activity effectively. Someone must put an

[226] For a discussion of the informational self see Belliger/Krieger (2018).

[227] See Floridi (2005).

end to the discussion and give orders that others carry out. Throughout history, the much-discussed and not well-defined concept of "power" is always illustrated by hierarchies of one kind or another. Governmental bodies, businesses, educational institutions, religious communities, indeed, every form of cooperative endeavor among people is organized through power, which seems inevitably to flow from the top down in the form of command-and-control communication. It is no accident that the commands of God come from above.

Although for Luhmann (1975), power represents that symbolically generalized medium that is the basis of the functional subsystem of politics, he admits that power is constitutive of social order in general and, therefore, embedded in all communication. All communication, in one way or another, depends on a functional subsystem organized by power. The extent to which systems are necessarily hierarchically organized or what systemic hierarchy means must be answered regarding the systems theoretical constitutive principles of control and steering. An autonomous, self-referential, operationally and informationally closed system must be able to control its operations to maintain its autopoiesis. Control and steering is what agency means in systems theory. Agency manifests itself as autonomous, adaptive behavior. Networks, on the contrary, are not constructed with regard to any kind of self-referential operations and the maintenance of setpoints. Consequently, networks do not need a "controller." Networks are neither machines nor organisms. Networks do not operate to maintain the operations and organization of a limited set of elements. Networks know no homeostasis. Nonetheless, networks exhibit durability and resistance to change based on the quantity of links that must be differently translated and enrolled if the network is to be changed. We are not claiming that there is no such thing as power in network order. In ANT, power is usually understood in terms of constructive agency, which is the ability to be a mediator and construct information through translation and enrollment processes. Agency is an inherent characteristic of all actors. Even when actors cease to be mediators and become intermediaries in black boxes, they remain capable of regaining "power" to change the network. ANT is a theory of distributed agency, whereas systems theory must centralize control within a system and assume an ecological "balance of power" among structurally coupled autonomous systems.

Even if one does not go so far as ANT when decentralizing power, it is clear that for modernity and most of Western history, the pyramid could be seen as the best visualization of power.[228] It is a structure that can be found in the organogram of almost every organization in society. Therefore, popular resistance to power or attempts to gain power are always interpreted as bottom-up movements, whereas attempts to maintain power are always top-down. This assumption underlies what has come to be known as "critique" in the modern age. Critique is based on the belief that social communication has always been vertical; orders come from the top, whereas compliance or resistance comes from below. The digital transformation disrupts hierarchical power by allowing effective communication and cooperation not only from one-to-many but from many-to-many. The affordances of digital technologies encourage lateral and distributed communication. If social space, as Latour never tires of saying, is indeed "flat," then power cannot be hierarchical. As we will see below, it has become clear in today's globally networked society that traditional bureaucratic organizations are becoming increasingly dysfunctional and uncompetitive. Hierarchies are everywhere being replaced by network organizations based upon distributed decision-making and self-organization.[229]

Finally, the third disruption that characterizes the digital transformation is the transformation of the order of knowledge. Knowledge depends on media. If what we know cannot be communicated through some durable medium, it disappears with us and does not become part of society. Without media of some kind, there is no knowledge. Social and political revolutions have always accompanied revolutions in media. The invention of writing replaced oral tradition and changed society. The invention of the printing press and the electronic mass media changed society. The invention of digital media is currently changing

---

[228] For Latour (2013), power, at least political power, is best visualized as a circle and not a pyramid.

[229] See Belliger/Krieger (2016) for a discussion of networked organizations.

society. The so-called "new media" have introduced a new order of knowledge that is non-hierarchical, unlimited, connected, inclusive, complex, and open to everyone.[230] The new media are non-hierarchical because they give everyone the means of producing and distributing information. They are unlimited because the cost of producing and distributing information has drastically been reduced, thus eliminating the old economy of scarcity in information and knowledge. The new media are connected and, therefore, inclusive. Nobody, or at least very few, do not have access to all the information on the Web. It is almost impossible to block or limit access to information, as all the leaks, whistleblowers, hacks, and disclosures demonstrate. The new order of knowledge is complex since there are no longer gatekeepers, authorities, institutions, or trusted sources of knowledge. Knowledge comes from anywhere, indeed, everywhere, as the acceptance of so-called "citizen journalism" by mainstream media shows. And finally, the new order of knowledge is public in a way in which knowledge previously has never been. The very idea of a "public sphere" arose in the modern period because of media proliferation and the increasing availability of information. But the public, in the traditional sense of an arena in which citizens participate in democratic deliberation, has long become a global socio-sphere in which politics is only one of many forms of communication and participation.[231]

These three disruptions, the dissolution of the bounded individual, non-hierarchical network organizations, and the new order of knowledge that dismantles traditional roles and functions, are emerging from the affordances of digital information technologies and are creating a global network society. This contemporary society, or as Latour would say, the collective, can no longer be understood or regulated based on the values, norms, and forms of power typical of modern Western industrial society. New norms and new forms of organization and regulation are required if we are to move into the global future.

[230] See Weinberger (2012).

[231] For a discussion of ANT and new media see Krieger/Belliger (2014).

4.2 Network Norms

Just as the prehistoric stone ax placed certain constraints on what it could be used for and what it meant to become a hunter, warrior, or builder, so do the new information and communications technologies shaping our world have their own affordances. The symmetrical and distributed agency ANT proposes obliges us to understand technologies and artifacts as social partners contributing in their own ways to what and who we are and what kind of society we live in. We, therefore, assume that digital technologies shape and condition what we can do and who we can become in the actor-networks they allow. Shaping and conditioning are not to be understood as irresistible forces or determining constraints but as normative guidelines. Affordances have a normative character. They influence, suggest, encourage, dispose, and nudge but do not determine. They guide behavior and experience. It can, therefore, be said that they condition the kind of beings we are and the society we live in. Since, as noted above, all construction is subject to the question of value, of whether what is being constructed is being constructed well or badly, the guidelines derived from the affordances of digital technologies can be considered normative. The network norms describe what it means to construct something well in the digital age. In this sense, one could also understand them as the normative framework of the global network society. In other words, if networking is the game, both human and nonhuman actors are the players, and there are rules in the form of normative guidelines about how to play the game well. The most important of these guidelines, which we propose to call network norms, are listed briefly below.[232]

4.2.1 Connectivity

Networking tends to connect everything to everything. This is generally the case, no matter what technology is being used. As soon as the stone became an ax and the hominin became a hunter or a warrior, the network

[232] This list is neither exhaustive nor exclusive. More or other norms, just as with Latour's modes of existence, can be described and most probably will be.

began to grow, expand, branch out, and scale up indefinitely. It is of the nature of meaning, the collective, to gather all things together. Whether we are wielding stone axes or programming computers, we are doing the same thing, only on a different scale and with different technologies. Networking in the digital age makes this inherent tendency of all networking explicit in a unique way. Indeed, it is probably no accident that a theory of network order has arisen at this point in history. In a sense, our technology is telling us who we are.

Generally, network connectivity is precisely the opposite of that which closed systems do. The "relationing" of elements, which, along with selection and steering, is one of the foundations of systemic order, is not the same as connectivity. Where closed systems are characterized by hierarchy, limitation, exclusion, and reduction, networks are non-hierarchical, inclusive, connected, complex, and public. On the most fundamental theoretical level, this follows from the assumption that Being is information and information is relation, not substance. To be, in the perspective of ANT, is to be related. Therefore, the inherent dynamic of networks is for everything to become ever more related, to gather more associations, and to integrate more actors in more complex narrative scripts. Whereas systems arise as attempts to reduce complexity, networks arise to increase the number of associations, relations, links, and connections between things. Whereas there can be no system of everything, networks strive to become universally inclusive. Whereas the basic principle of systemic order is exclusion, the basic principle of network order is inclusion.

Connectivity is the network norm that accounts for the necessarily global nature of network society. Luhmann also emphasized the global nature of the functional subsystems, but only at the cost of excluding everything that did not fit into the constitutive binary codes of these systems. Systemic order is indeed universally inclusive, but only because it excludes everything outside the system's boundary. The economic system, for example, extends globally, but only with respect to things that can be bought or sold. The legal system extends globally, but only concerning what can be judged as legal or illegal. The binary coding of the functional subsystems is an inclusion/exclusion mechanism. Networks know no such principle of exclusion and are not coded or organized by such binary distinctions. Networks are not only not territorially bounded, as is Luhmann's society, but they are also not differentiated based on inclusion/exclusion codes. Networks, of course, do have boundaries, but such boundaries are accidental and not essential and are, therefore, not constitutive of the network. The networks' borders are fuzzy and always provisional, open to renegotiation, constantly changing, porous, and, as we shall see below, flexible.

It is important to note that connectivity is not a weakness but a strength. It is not a bug but a feature. Let us recall that Paul Barron's original concept of a distributed network – which became the basis of the Internet – was a solution to a security problem. It was the only form of order that was designed to withstand a nuclear attack. All those who equate security with impenetrable boundaries of some kind, whether it be city walls, motes around castles, closely guarded borders, or even The Great Wall, do not understand that the distributed network is a security strategy. Contrary to systemic order, which is founded on drawing clear boundaries, building, so to speak, walls around systems, networks are permeable and base their strength on scalability, decentralization, openness, and flexibility.

The more extensive the network, the more connections it has, the more resilient it is. ANT has an interesting and suggestive correlation between reality and the quantity of associations in a network. The more links a network has gathered, the more "real" one can say the network is. This follows directly from the relational ontology of ANT. If Being consists of relations, the more connections there are in a network, the more effort it takes to change the program of action of the network, the more resistance the network offers to change, and thus the more durable and "real," it is. When the network is large, there are many more nodes and links that must be moved. When the network is highly connected, it requires more effort to divert its trajectory or subvert its program of action. Concerning the digital transformation, it has become apparent that the so-called "virtual" world is becoming more real and effective than what was previously thought to be solid, physical reality.[233]

4.2.2 Flow

A second network norm can be termed "flow." Networks allow, encourage, and predispose information to flow freely through all the nodes. Flow is not merely passage or movement like water flows through a pipe. As a network norm, flow is derived from the fact that all actors in a network are, in principle, not mere intermediaries through which information moves but mediators who can change how information in the network is being constructed. Therefore, this notion of flow is not to be equated with Latour's distinction between the setup of a network and what flows through it. When actors are mediators and not mere intermediaries, they change not only what flows but the setup of the network. Flow in the sense of the word that we intend means that information and everything else, such as people, money, goods, etc., are always uncontrollably and unpredictably making associations and configuring and reconfiguring the network. Technical mediation, processes of translation and enrollment, construct information. Information implies that there will always be surprises, there will always be unforeseen, uncontrolled, unchanneled associations, unexpected relations, and disruptive connections. Unlike the carefully limited and controlled communication within a functional subsystem, networks do not know, cannot know, and cannot control information construction. No more than our prehistoric hominin could tell in advance what the stone would tell it about using an ax and becoming a hunter, so we today cannot in advance know what AI will mean for human existence and society. Regarding the construction of information, we are only one player in the game that all beings are playing. As we shall see below, there is ample evidence that digital technologies have disrupted traditional forms of information control. Despite untiring calls for regulation and a return to centralization, gatekeeping, censoring, etc., it has become clear that attempts to control or limit information flows weaken

[233] For a discussion of the "mixed reality" of digitally mediated society see Krieger/Belliger (2014).

the network or invite workarounds and extensions into other networks – the darknet is a case in point.[234]

4.2.3 Participation

As mentioned above, all actors in a network are potential mediators, that is, agents of information construction. Participation as a network norm means that all nodes in the network are not mere intermediaries or passageways through which information – or anything else – moves, but every node also has the ability, and even the duty, to change, improve, transform, and repurpose information. Every node in the network is an actor, a source of information, a contributor to the whole, and not a mere function, a cog in the machine. Participation is what distinguishes the functional elements of a system from actors in a network. Participation makes not only the flow but also the content of information unpredictable and uncontrollable. Participation is agency. It could be considered as the normative force of Being, or one could say, the "obligation" to exist, to actively construct meaning. As Shakespeare put it, it is the question to be or not to be. The answer to this question is participation.

4.2.4 Transparency

Transparency means that the sources, reliability, and uses of information are known by all. Nothing is hidden; this is the famous "glass human being." Publicy, not privacy, is the default condition today.[235] Transparency, however, must be symmetrical. The network does not condone or facilitate asymmetrical transparency. The futility of attempts to maintain asymmetrical transparency is what all the leaks, hacks, disclosures, and whistle-blowing teach us. The network norms of connectivity, flow, and participation oblige that everyone and everything be equally transparent. Transparency is the opposite of anonymity. Those actors

[234] Another case in point is the current proliferation of AI and the many haphazard attempts by regulators, both private and public, to keep AI under control.

[235] For a discussion of "publicy" as the new norm replacing privacy see Belliger/Krieger (2018).

who depend upon anonymity, such as criminals or state actors, and all the spies, bad actors, trolls, etc., whatever their interests and whoever they serve, are contributing to the general insecurity of the network and not making the world safer. Many of the dangers and harms of digital technologies, including recently AI, can be traced back to anonymity, disguise, deception, and strategies to avoid identification and accountability. Lack of transparency fosters not only opportunities for misuse of technologies in all forms but also fosters mistrust. Transparency means that networks are based on trust. It is no secret that secrecy has always been a means of power, advantage, inequality, and misuse. The digital transformation and the advent of a global network society present the opportunity to develop new forms of regulation other than those arising from traditional power asymmetries that depend on privileged access to and control of information.

4.2.5 Authenticity

Authenticity is related to transparency but designates the network affordance of truth. It is commonplace that networks function based on trust. In networked organizations, mistrust is dysfunctional. In the Machiavellian world of bureaucratic hierarchies and informational scarcity, mistrust is normal and perhaps even necessary for survival. This is no longer the case in connected, participatory, and transparent networked organizations. Authenticity takes up but transforms the modern Western ideal of the free, self-determined individual, the autonomous, rational subject. Against Western individualism, the network norm of authenticity does not mean that the individual must free itself from the pressures of social conformity. Authenticity does not mean that one must strive to discover one's true self or that the true self is an individual. It does not mean one must become the architect of one's destiny. It does not subscribe to any notion of human nature, freedom, or self-determination. To say that networking obliges one to be authentic means that individuality is experienced as a specific kind of agency, such as mediating associations, constructing information, and becoming networked. Authenticity means that one accepts that the actor is the network and that agency should always be in the service of gathering the collective or, in other words, playing the game of meaning well.

4.2.6 Flexibility

Networks are never stable, even if they may appear so. Unlike closed systems, open networks do not consist of elements subordinated to clearly defined functions within a whole. The table system constructs the legs of a table for one purpose. The system maintains its organization, integrity, identity, and autopoiesis only so long as all elements are related to each other in such a way as to ensure that the goals of the system can be attained. On the contrary, every "actor" in a network can forge new links, change old ones, assume different roles, and either expand or retract the network. Networks do not have clear borders that define what they are and what they do. They do not have fixed functions and roles, as do the elements of a system. Networks are open to change and redesign in many directions. They often integrate different identities and purposes. Flexibility means that innovation and transformation are more valued than sustainability and fixed programs of action. It is not what you are but what you can become that guides and motivates action. As a network norm, flexibility replaces sustainability, a systemic concept based on maintaining a system's specific values, constants, or setpoints. What sustainability means for systems, flexibility means for networks.

4.3 From Systems to Networks

If the above-listed network norms do influence how the networks that we live in are constructed, this should be visible. There should be apparent differences between traditional forms of organizing cooperative action in society and the forms of organizing emerging based on networking. The first two parts of this book were dedicated to theory. We now turn to empirical reality. The question is no longer which scientific paradigm is more coherent, heuristically powerful, universal, or explanatory. We now ask: What does society look like if and when systems become networks? In the following, we will briefly examine some of the more significant trends that characterize organizing in some of the major areas of the global network society. These are the areas of business, education, and healthcare. We assert that the trends becoming apparent in these areas are also typical for other areas of society. Be that as it may, these examples should be sufficient to illustrate how networking is shaping society today. Although the examples we have chosen are neither exhaustive nor even representative, we hope that the reader can understand how the network norms briefly described above are concretely influencing society. They are intended to illustrate, even if only superficially, how systems today are becoming networks. Leaving the terrain of pure theory and venturing into empirical description has solely the purpose of a modest illustration. The discussion below is not meant to be empirical sociology but merely to show how the digital transformation is ushering in a global network society.

4.3.1 Networks Do Not Define Fixed Roles and Functions

Networking in Business

Let us begin by focusing on business. The influence of the network norms on organizing in the private sector and beyond can be seen in the many new management practices that disrupt traditional organizational roles and functions. This is the case today, even for not explicitly networked organizations.[236] A new understanding of leadership and new employee roles are being implemented in organizations of all kinds, both production and services. Before command-and-control communication was standard, employees on all levels now contribute to and influence decision-making. One speaks of "servant leadership"[237]

[236] An organization can be considered "networked" when informal, non-hierarchical structures have precedence over traditional bureaucracy, high clustering and short paths of communication, connectivity and opportunities for random encounters, and context steering. See https://hbr.org/2015/06/what-makes-an-organization-networked, and Wikipedia https://en.wikipedia.org/wiki/Network-centric_organization.

[237] See Wikipedia https://en.wikipedia.org/wiki/Servant_leadership.

to designate new forms of management that are increasingly replacing traditional management. Furthermore, employee roles are becoming more participative and self-directed. Wüthrich et al. (2009) and Kaduk et al. (2015) document and describe how managers today are called upon to "break patterns" that have long been defined as best practices. Notably, they do not attempt to describe networked organizations but successful management practices in today's world, regardless of what kind of organization. The organizations cited as examples of "pattern-breaking management" are not at all the typical examples cited in the literature, that is, high-tech and internet companies. These examples show that breaking traditional roles of industrial management is imperative for management in all kinds of organizations today, even traditional businesses.

The typical management patterns in the industrial age were control, standardization, rational decision-making, optimizing for short-term gains, efficiency, command and control communication, well-defined and routine processes, predefined structures, clear organizational boundaries, risk avoidance, compliance, and clearly defined job specifications. All these management practices can be modeled in terms of the selection, relationing, and control functions characteristic of systems. Indeed, systems management arose almost simultaneously with cybernetics and has been part of systems science ever since.[238] Faced with the growing complexity and dynamic of today's VUCA world, many of these traditional pillars of management wisdom and practice have dissolved into paradoxes and contradictions.[239] Formulated as paradoxes, these new trends in management could be described as attempts to steer unsteerability, to mistrust trusted control mechanisms, to standardize diversity and innovation, to shortsightedly see

---

[238] See for example Boulding (1956); Beer (1959); Kast/Rosenzweig (1972). See also Wikipedia https://en.wikipedia.org/wiki/Management_cybernetics.

[239] There is a vast knowledge management literature which documents these trends. See for an overview Wikipedia https://en.wikipedia.org/wiki/Knowledge_management. VUCA stands for volatility, uncertainty, complexity, and ambiguity and was coined in 1987 based on the leadership theories of Bennis and Nanus to describe or to describe the conditions under which organizing in today's world must be done. See Wikipedia, https://en.wikipedia.org/wiki/VUCA.

into the future, to decelerate acceleration, to know the unknown, to risk risk-tolerance, and to freely choose and vary constraints. Where once order and stability were valued, now organizations are striving to be resilient on the edge of chaos. Once command and control communication, hierarchy, clearly defined processes, and structures were valued, organizations now attempt to foster knowledge networks, teams, interconnectedness, open communication, collaboration, and transparency. Management is no longer able to rely upon strategies for making things simple. It is a fundamental principle of systems theory that complexity cannot be managed with simplicity but only with more complexity. What we are witnessing today, however, is that at a certain point, the increase of internal complexity in a system transforms the system into a network. The increasing internal complexity of many contemporary organizations at once dissociates and decentralizes teams and sections and increases their independent contact with the environment. In addition, changing roles for consumers who, via the widespread adaptation of "agile" methods for product development and independent access to information, have become "prosumers" tend to erase the fixed roles and functions in organizations of all kinds.[240]

The tendency of organizations to shift from systems to networks is nowhere more apparent than in the omnipresent trend toward *decentralization*. Decentralization is a management strategy that has become a model for managing networked organizations. In response to the challenges of an ever more complex environment, systems react by building up more and more internal complexity. This is done by decentralizing decision-making. Large organizations delegate more autonomy and independence to smaller working units. This fosters trust, enables flexibility, brings decision-making closer to the practitioners and the customers, and motivates employees at all levels to take responsibility for the whole organization. Traditional management roles are transformed thereby into what is known as "servant leadership."[241] Leaders

---

240 Agil methods for project management began in the software industry and have since spread to all sectors. See Wikipedia https://en.wikipedia.org/wiki/Agile_software_development.

241 For an overview see Wikipedia https://en.wikipedia.org/wiki/Servant_leadership

become more of a support or a coach for the smaller units instead of using budgeting and controlling to dictate what should be done on lower levels. Fostering open communication, trust, transparency, and self-organization are becoming acknowledged and even required management skills. The end of this process is marked by the transformation of the closed system into an open network.

Other important pattern-breaking management strategies could be self-management, cooperative, or bottom-up management. Wüthrich et al. cite the well-known Orpheus Chamber Orchestra that operates without a conductor in that it "rotates musical leadership roles for each work and strives to perform diverse repertoire through collaboration and open dialogue."[242] Bottom-up decisions about roles and functions are a typical scenario for Holocracy and Sociocracy, where workers choose their own roles and can flexibly change positions, and there is no longer a CEO.[243] The guiding principles of this kind of management are to give power to those who do the work, encourage personal responsibility, enable flexible role definitions, distribute decision-making, support cooperation, and promote a culture of collaboration, listening as well as speaking, valuing consensus, and devoting oneself to the task at hand. A further example is the Brazilian city of Curitiba, which managed to deal with many of the typical urban problems, such as unemployment, health care, education, environmental pollution, transportation bottlenecks, etc., through encouraging and enabling citizen participation.[244] What is significant about Curitiba is that instead of top-down projects, the city government supported the initiative and self-responsibility of its citizens to find and implement creative solutions. In this way, planning and decision-making became a social process enacted in non-hierarchical, inter-organizational, and networked communication.

A further management strategy that is typical of networking is trust-

---

242 According to Wikipedia https://de.wikipedia.org/wiki/Orpheus_Chamber_Orchestra.

243 See Wikipedia https://en.wikipedia.org/wiki/Holacracy; and https://www.sociocracyforall.org/sociocracy/.

244 See Wikipedia https://en.wikipedia.org/wiki/Curitiba.

based leadership. Hierarchical organizations are characterized by mistrust and rely extensively on bureaucratic rules, rigid systems of control, and discipline. Transparency and open communication are not characteristics of traditional management. On the contrary, trust-based leadership enables all co-workers to access information about the entire organization. Furthermore, all employees are encouraged to participate in contributing new ideas and participating in making decisions. These are issues usually discussed under the rubric of organizational culture. Trust, transparency, participation, open communication, and information sharing are values and attitudes, not merely processes. They are based on the network norms of flow, participation, authenticity, and transparency and have become, in many organizations today, more or less explicit expectations. Norms, however, should not be confused with standards and standardization. Traditionally, management was concerned with setting up standardized processes, job routines, measurement criteria and methods, qualifications, assessments, incentive programs, and other instruments to allow for prediction and control.

Wüthrich and Kaduk cite management strategies in human relations and human development in which the usual laid-out career paths with indicators and milestones are explicitly avoided. Instead, individual preferences, personal interests, abilities, an open horizon of possible paths to take within an organization, self-assessment, and similar options replace standard qualifications and assessments. Where once there was uniformity and standardization, now there is an emphasis on diversity, heterogeneity, and innovation. It has become almost commonplace in management theory to emphasize the importance of diversity. Heterogeneous teams, for example, have demonstrated better problem-solving results than groups with similar backgrounds, experience, and knowledge. This is a cornerstone of Surowiecki's *Wisdom of the Crowds*. Surowiecki (2004) showed many examples of how heterogeneous groups are more effective at problem-solving than experts with similar backgrounds. In general, it can be argued that the participatory culture of networking fosters diversity because it normalizes the abnormal and gives equal opportunities for expression to all. In strategic management, Wüthrich and his colleagues point out the importance of flexibility and innovation. Flexibility and an emphasis on innovation foster and support resilience. The network norm of flexibility allows for unforeseen goals and avoids fixed setpoints. Organizing in such a way as to enable many possible goals or programs of action, the free flow of information, and encouragement of flexibility and innovation is not only to be found in the business world. Organizing based on openness to changing roles and functions can also be seen in education.

Networking in Education

The advent of networking as the most important form of constructing social order has become apparent not only in business but also in education. The influence of networking has been documented by early studies from The World Bank (2003). Collis (2005) has noted that in all nations, education must create a workforce that can "derive local value from information often in creative ways that go beyond expected performances," that is able "to work in multidisciplinary and distributed teams," that can "use information technology (IT) for knowledge management, sharing, and creation," and that can meet "the need to 'act autonomously and reflectively; joining and functioning in social heterogeneous groups'" (Collis 2005: 215). Digital technologies have brought forth an entire pallet of education tools and methods that are disrupting traditional educational practices.[245] Among others, these are e-learning, knowledge networks and communities of practice, open educational resources (OER), learner-centered instruction, mobile learning, social learning or learning 2.0, workplace learning, MOOCs, learning analytics, big data in education and AI, personal learning environments (PLE), partnerships among educational institutions, flipped classroom, and connectivism as a new didactic and pedagogical theory.

[245] There are many trend reports for education. See for example the Horizon Reports http://www.nmc.org/publication-type/horizon-report/ , or the OECD Talis Report http://www.oecd.org/edu/school/talis.htm , or the European Schoolnet Briefing Papers http://www.eun.org/observatory/surveyofschools , or the Partnership for 21st Century Learning http://www.p21.org/ , or the Education section of the Digital Agenda for Europe http://ec.europa.eu/digital-agenda/en/ict-education, and in the area of learning and development the Cegos Study 2015 https://www.integrata.de/fileadmin/Bilder/Service/Downloads/2015_CEGOS_Research_-_LD_Globalization.Final_Report.pdf.

This list is not exhaustive and perhaps not even representative since new media in education, or technology-assisted education, or whatever word or phrase is being currently used cannot possibly address the entire scope of theories, practices, experiments, pilot projects, strategies, programs, initiatives, etc. that characterize the digital transformation of education on all levels and in both private and public sectors today. The administration of educational organizations has much the same tasks as management in any organization. Organizational development, personal management, finances, marketing, quality management, controlling, etc., are similar in most organizations. The affordances of ICTs in the global network society demand educational decision-makers to break out of the patterns that have guided education for a hundred years or more. Decentralization and the practices of self-management, trust-based management, and bottom-up management are also to be found in education. One will find many of the new functions and roles for management that we have already noted for business in educational administration as well. However, curriculum development, course administration, assessments, knowledge transfer, and learning evaluation are specific to education.

Applied to education, these norms do not merely change traditional administrative roles, but instructors and teachers are becoming coaches who guide the self-directed learning processes of students and trainees. New forms of curricula such as MOOCs, social and mobile learning, open educational resources, learning analytics, and personal learning environments, to name only a few, transform the traditional roles of both teachers and students. We will discuss these new educational developments below from the perspective of how networking pushes educational institutions and organizations beyond conventional boundaries and forms of management.

Networking in Healthcare

The digital transformation of healthcare from the perspective of the disruption of traditional roles and functions can perhaps most clearly be seen in what has come to be known as the e-patient movement.[246] The e-patient movement is formally organized around the non-profit Society for Participatory Medicine.[247] The e-patient movement defines itself as "a movement in which networked patients shift from mere passengers to responsible drivers of their health."[248] In contrast to the Quantified Self movement, which we will discuss below and operates mainly in the secondary healthcare sector of fitness and prevention, the e-patient movement operates within the primary healthcare sector of diagnosis, therapy, and rehabilitation. A new role for patients and new relationships between patients and healthcare providers such as doctors, clinics, hospitals, insurers, regulative agencies, pharmaceutical companies, and the medical technology industry characterizes networked healthcare in the primary sector. The driving force in this new constellation of roles and relationships are the patients, who, based on newly gained access to medical information, increasing demand to play a more participatory role in healthcare, and connection to other patients are becoming more active and self-determined. The explicit goals of the Society for Participatory Medicine are: 1) "to guide patients and caregivers to be actively engaged in their health and health care experiences;" 2) "to guide health professional practices where patient experience and contribution is an integral goal of excellence;" and 3) "to encourage mutual collaboration among patients, health professionals, caregivers, and others allowing them to partner in determining care."[249]

Examples of networked communities are Patients Like Me,[250] CureTogether,[251] and Acor.[252] In these communities, patients are linked to peers, information throughout the Web, and medical professionals. Not only are patients connected to each other and healthcare professionals, but the flow of information, including access to medical research,

246 In general see Wikipedia https://en.wikipedia.org/wiki/E-patient.

247 http://participatorymedicine.org/.

248 http://participatorymedicine.org/about/.

249 http://participatorymedicine.org/about/.

250 https://www.patientslikeme.com/.

251 http://curetogether.com/.

252 http://www.acor.org/.

makes it possible for some patients to become better informed about their specific diseases than many doctors. Another example of changing roles and functions is that patients can publicly rate the performance of medical service providers. CureTogether.com, for example, offers access to millions of ratings not only of doctors and hospitals but also treatments and medications.[253] Regarding participation, e-patients can donate their data to medical research, whereas doctors share data via such programs as "Open Notes."[254] An example of data donation that the US government supports is the Blue Button Movement. The Blue Botton makes it easy for patients to access their medical records.[255] Another example is 23andMe, a web-based personal genome service, which allows individuals to get information about their genetic disposition for certain diseases and donate this data to research.[256] Echoing the explicit goal of the Society for Participatory Medicine in the USA, "letting patients help," the EU Action Plan 2012-2020 for using digital solutions in healthcare has the slogan: "Putting patients in the driving seat: A digital future for healthcare." The traditional healthcare system did not allow, encourage, and enable patients to play an active and constructive role in medicine. The shift from systems to networks in healthcare changes this situation radically. Healthcare networks empower patients to responsibly and effectively use their "health literacy" and new networked technologies in ways that were utterly unforeseen in traditional healthcare.[257]

A further example of changing roles in medicine is the practice of "shared decision-making." In shared decision-making, the interaction between doctors and patients becomes cooperation that is based on access to shared information and the explicit acceptance of equal decision-making rights. Diagnosis and therapy cease to be a one-way street

[253] http://curetogether.com/.

[254] http://www.opennotes.org/.

[255] https://www.healthit.gov/patients-families/join-blue-button-movement.

[256] https://www.23andme.com/en-int/.

[257] "Health literacy is the ability to obtain, read, understand and use healthcare information to make appropriate health decisions and follow instructions for treatment." https://en.wikipedia.org/wiki/Health_literacy. See Bessiere et al. (2010) for a discussion of the positive effects of networked health.

and become a collaborative endeavor in which doctors and patients work together as a team. Networking allows patients to gain access to information, medical records, peer communities, and other resources that were not available before networked forms of healthcare arose. Furthermore, e-patients can now contribute to research by donating data and collaborating with scientists on agenda-setting for clinical studies and research programs.

These examples show that connectivity, flow, communication, and participation have enabled normal people to access, understand, and use health-related information in new and unforeseen ways. This changes patients' expectations toward healthcare professionals and institutions, as well as the possibilities that these established players have in dealing with their "customers" and "clientele." Patients and caregivers can now easily obtain information on all aspects of their health concerns, communicate, and get advice and support from others in similar situations and healthcare professionals. Patients can participate constructively in diagnosis, therapy, and rehabilitation. As in other areas of society, healthcare is becoming a part of the "sharing economy."[258] Challenges for all those who take on new roles in healthcare lie in managing the networks that are emerging from the digital transformation in such a way that costs are lowered, the effectiveness of interventions and prevention is increased, communication between doctors and patients is improved, and extended, medicine becomes more personalized, and transparency and trust replace hierarchical relationships. In short, one could say that the democratization of healthcare is an effect of networking.

4.3.2 Networks Have No Constitutive Boundaries

Business

One of the most significant changes that is taking place because of the digital transformation is the breaking down of institutional and organizational boundaries. In comparison with systems, networks do not have constitutive boundaries. All of the functional subsystems of

[258] For an overview see Wikipedia https://en.wikipedia.org/wiki/Sharing_economy.

society that Luhmann describes are constituted as systems by a specific binary coding such that communications can be ascribed to one system or another and processed accordingly. Business, for example, is coded by buying/not-buying (paying/not-paying). If you go to the market and don't buy anything, there has been no financial transaction; therefore, whatever you do at the market is not business. The same is true for education, which is coded by the binary difference of certification/non-certification. If you are an autodidact without any means or intention of certifying your knowledge and skills, the learning you are doing is not part of the educational system. Healthcare is also a functional subsystem of society, constituted by the difference between sick/well. If you are well, you do not participate in the healthcare system. No insurance will cover the costs of your wellness, and no doctor will treat you for it. When systems become networks, these constitutive boundaries become porous and diffuse.

Networked organizations are constituted not by binary rules of inclusion/exclusion and clear boundaries but by attempts to establish and maintain various cooperative relations. If one looks at the problem of competitive advantage in a global market facing many companies today, closed systems are not well-positioned for success. Lavie (2008) notes that the "emergence of interconnected firms is a contemporary phenomenon" (325) rooted in the economic realities of the global network society. To explain why networked firms have competitive advantages, he introduces the notion of "network resources" (325). "Network resources reside in alliances in which the interconnected firm is involved rather than within the scope of the firm's organizational boundaries" (2008:325). Alliances serve various goals. They allow "flexible and more rapid adjustment to changing market conditions," "reduce time to market in response to shortened product life cycles," "assist in bridging national boundaries," and "reduce market uncertainty and stabilize the firm's competitive environment by forming norms of reciprocity that establish commitment and regulate exchange transactions" (325). Furthermore, alliances are not restricted to any one branch or sector. They have proven efficient in research and development, production, and marketing "in almost any industry" (326).

As Saz-Carranza and colleagues pointed out, and for Lavie as well, the networked firm is involved in associations that are "dynamically evolving yet durable" (2008: 326).

Whereas traditional theories of the firm emphasize interfirm competition, theories of networked organizations emphasize alliances, partnerships, strategies for managing associations, inter-organizational governance structures, organizational learning, or knowledge management, and they focus on the achievements of the entire network. This new networked understanding of organizing departs from traditional views that see the appropriation of value as exclusively based upon ownership or disposal over resources such as capital, raw materials, and labor. Against this conventional view, Lavie points out that for networked organizations, there is a "direct sharing of resources" and an "indirect transferability of benefits associated with these resources" (328). He concludes that the "proprietary assumption of the resource-based view prevents an accurate evaluation of an interconnected firm's competitive advantage" (328). Networked forms of organization can create and appropriate value more efficiently and competitively because of their ability "to maintain valuable interaction with their partners" (332) and less because of their inner-organizational assets. Lavie concludes:

> At the turn of the 21st century, alliances have emerged as a primary vehicle for conducting economic transactions. Once conceived of as independent entities, firms are now considered interconnected in the sense that they engage in multiple alliances with counterpart firms. This phenomenon necessitates a new theory of the firm that incorporates simultaneous competition and collaboration as drivers of value creation and appropriation. […] Managers need to pay attention not only to the value of resources nurtured by their own firms but also to the resources possessed by their firms' partners. They must evaluate the value of network resources as well as the value of potential combinations of network resources and internal resources. They must collaborate to create value and compete to appropriate their relative share of that value. They must look

> beyond the immediate scope of alliances and seek to leverage network resources while finding the right balance between resource sharing and the protection of proprietary assets. (Lavie 2008:333)

Understanding the advantages of networked organizing from the point of view of transcending traditional boundaries and cooperatively using resources also explains the emergence of *global projects* as an important way networked organizations operate today. Ainamo (2008) claims that "a new 'logic of organizing' has been spreading in market economies" (482). This new organizational logic is based on "flattening out organizational hierarchies, weakening of firms' boundaries in favor of networks of collaborations, and restructuring of competition between firms within and across industries" (482). This has led to "global projects as new organizational forms" (482). Ainamo notes that "more and more companies are now becoming structured around distinctly global projects" (483).

> Organizing highly skilled workers dealing with complex problems to create novel outputs by integrating varied forms of expertise – also across cultures and institutional systems – represents a significant new organizational form. (483)

Barney (1991:101) has defined resources as "all assets, capabilities, organizational processes, firm attributes, information, knowledge, etc., controlled by the firm." According to Lavie (2008), sharing resources adds value and competitive advantage. From this perspective, global projects can be seen as a characteristic phenomenon of the global network society. Ainamo comments upon this situation by claiming that in today's world, "firms are transforming into fluid, overlapping organizational arrangements in which both internal and external boundaries break down, and cooperation with other firms grow" (482). Cooperative activities between organizations can take many forms. Instead of alliances or partnerships, projects have emerged as a common form of inter-organizational cooperation. Castells (2005) also emphasizes the project-based dynamic of networking:

> […] large corporations, and their ancillary networks, engage in strategic partnerships on various projects concerning products, processes, markets, functions, resources, each one of these projects being specific, and thus building a specific network around such a project, so that at the end of the project, the network dissolves and its components form other networks around other projects. (Castells 2005:9)

Projects, of course, are not organizations. They are usually distinguished from organizations in that they are directed to specific goals to be attained within a limited time and with limited resources. Nonetheless, projects must be managed by organizations. Typical project management tasks are initiating, planning, executing, controlling, monitoring, and closing the project.[259] Even though such activities usually take place within organizations, in many projects from industries such as the film industry, the software industry, the construction industry, biotech, research consortia, and consulting and marketing, it is often the case that experts from outside any one organization or from different organizations are explicitly contracted for certain projects. They work together on specific tasks whose completion also ends the contractual cooperation.[260] Ainamo has identified two ways organizations extend their borders to form projects. Some organizations can operate as agencies specialized in carrying out projects within a specific area for other organizations. The experts on these projects remain within the agency but move from one project to another or even work simultaneously on several projects in various organizations. Organizations can also cooperate or contract with independent external experts. The project is a heterogeneous or hybrid form of organization that consists of flexible, ad hoc teams that are free to regroup or move on to other projects once any single project is completed. Along with such global networked projects, there comes a transformation of what work means and what

---

259 See the Guide to the Project Management Body of Knowledge for a representative statement https://en.wikipedia.org/wiki/Project_Management_Body_of_Knowledge.

260 See the discussion of various project-based industries in Ainamo (2008).

demands are being made upon labor.[261]

It is an often-noted characteristic of labor in the knowledge society that knowledge workers, who traditionally were regularly employed by one organization, become independent contractors with the companies they once worked for and other companies, including competitors. Castells (2000: 12) distinguishes between "self-programmable" and "generic" labor. According to Castells, "self-programmable labor is equipped with the ability to retrain itself and adapt to new tasks, processes, and sources of information, as technology, demand, and management speed up their rate of change" (12). Labor is no longer an organizational role or something that only has its place within the boundaries of an organization. Knowledge workers are experts who often make up the flexible and heterogeneous project teams responsible for business activities in many areas today.[262] What project forms of organizing show is that flexibility with regard to organizational boundaries has become imperative for both employers and employees. The major advantage of project-based organizing is adapting and reconfiguring operations, participants, goals, and processes quickly and globally. Of course, not everyone is a knowledge worker and finds themselves able to profit from the freedom implied by variable and open networks. For those whom Castells puts into the category of "generic" labor and those who fall out of any form of employability, new networked forms of education and civil society can help alleviate many problems.

Another important example of networks going beyond the boundaries of traditional organizations is platforms. Everyone knows how Microsoft, in the early years, took over the PC Market from Apple. Apple's products were locked into a proprietary pipeline, a one-sided value chain in which consumers were offered products from a producer. Microsoft changed the rules of the market by making their operating system open

---

[261] We will return to this aspect of the network society when discussing challenges for education below.

[262] Castells (2000) speaks of the "individualization of labour" as the key transformative factor and characterizes this workforce as, "part-time work, temporary work, self-employment, work by contract, informal or semi-formal labour arrangements, and relentless occupational mobility" (12).

for all to use as a platform upon which independent producers could offer their products, whether hardware or software. Microsoft quickly dominated the PC market, and Apple was on the brink of bankruptcy. Apple, of course, learned the lesson. When the iPhone came onto the market, Nokia, Motorola, and other manufacturers of telephones were dominant. The iPhone, however, was not another telephone; it was a platform that networked app developers and consumers like Microsoft did with Windows. Google soon followed with Android. Today, there are other examples of platforms such as Airbnb, eBay, Uber, and Amazon, not to mention the many innovative companies from China. These new forms of doing business are *networked markets*. Like traditional markets, they bring buyers and sellers together. But platforms bring many sellers and buyers in such a way together that all profit from synergy, growth, innovation, and exploration of mutual opportunities. This extraordinary dynamic is called "network effects."[263]

Various kinds of network effects have been documented. One such effect of networking occurs when an increase in product usage leads to a rise in product value. For example, the more people who use an iPhone, the more app developers are interested in creating apps for the iPhone, and the more valuable the iPhone becomes for users. Facebook is another example. The more people use Facebook, the more people it attracts since everyone is there. This is a positive feedback loop and is called a direct network effect.[264] There are also indirect network effects, such as when the increased usage of a product leads to innovation and production of complementary or associated products. Again, Microsoft is a good example insofar as the Windows platform, just as with Google Android later, led to the development of many compatible products. Another network effect can be called "two-sided" because the developer/producer and the consumer gain the value. The iPhone becomes more valuable for consumers because it creates value for developers. Uber or Airbnb can be seen as platforms with two-sided network effects.

---

263 See Liebowitz http://wwwpub.utdallas.edu/~liebowit/palgrave/network.html and the discussion and list of literature by Arun Sundararajan on network effects see http://oz.stern.nyu.edu/io/network.html

264 See the classic work by Katz and Shapiro (1985; 1994).

Uber creates value for both drivers and passengers. It can even be seen as a "multi-sided" platform because it generates opportunities for new services such as the delivery of food, doctors' visits, etc.[265]

Parker et al. (2016) have recently summarized some of the disruptions of traditional organizations introduced by platforms. In contrast to traditional organizations, platforms do not produce and market particular goods and services. They enable an entire "ecosystem" consisting not only of producers but also prosumers who are connected so that all gain value from the free flow of information, open communication, low barriers to participation, transparency, authenticity, and flexibility. Whatever products and services are being offered, platforms are organized as networks and not as markets. They focus on the synergies, cooperations, connectivity, and the optimal uses of generated data. Data is a resource that is not primarily based on ownership but on use. Sharing data in networks creates value for all participants. The only thing that needs to be organized and managed are the transactions. The participants contribute resources themselves. The major advantage of the network is the connectivity, flow, and participation of all stakeholders.

Education

Educational institutions are also benefiting from partnerships and collaborative communities of practice that go beyond traditional boundaries regarding problems and opportunities in the areas of infrastructure and pedagogy. Schools and educational institutions on all levels, such as primary or secondary schools, colleges, and universities, are connecting and building networks. They enter into network-type cooperations and communities of practice to exchange knowledge and best practices. This is taking place in both administration and teaching. Schools cooperate to develop better teaching materials, infrastructural solutions, and similar "products." In addition to sharing knowledge and developing new solutions, educational institutions enter into part-

[265] See the many articles on platforms and network economics by Economides http://www.stern.nyu.edu/networks/.

nerships to carry out projects to distribute and implement ideas and solutions beyond the participants in the network. Finally, such educational networks work with organizations such as businesses, museums, cultural organizations, etc., beyond the boundaries of their communities, cities, and regions. The goal is to make an entire community or region more attractive and competitive.[266]

Partnerships, networks, and new forms of inter-organizational learning, such as MOOCs, offer significant management challenges. The same can be said of Open Educational Resources (OER). We will discuss MOOCs below. Let us turn to Open Educational Resources (OER), which can be understood to illustrate the network norm of participation. New media studies (Jenkins et al. 2005) speak of "participatory culture." Forms of participatory culture are communities of practice, collaborative problem-solving, user-generated content, crowdsourcing, social networking, and many similar networked forms of collaborative work. In education, participation fosters peer-to-peer learning and a very untraditional attitude toward intellectual property. All participants produce information, which is placed in a commons (often under a Creative Commons license) so all can use it. Open Educational Resources are defined as

> […] teaching, learning, and research resources that reside in the public domain or have been released under an intellectual property license that permits their free use and re-purposing by others. Open educational resources include full courses, course materials, modules, textbooks, streaming videos, tests, software, and any other tools, materials, or techniques used to support access to knowledge. (Open Education Resources 2013)[267]

Properly connecting learners and learning resources is one of the most important tasks of educational management and teaching. As Jones (2015) points out, the affordances of ICTs "have the potential to change the relationship between learners and the institutions of learning to

---

[266] For a typology of school networks see Muijs et al. (2011:37ff.).

[267] See also https://en.wikipedia.org/wiki/Open_educational_resources; and Jones (2015).

their resources." This potential, however, can only be realized when it is "institutionally enacted" (120). This means that educational institutions or content providers "make their materials freely available online for anyone to use" (121). The usual way to do this is to license texts, images, videos, etc., under the Creative Commons license, which puts content into the public domain that does not stand under the usual copyright restrictions. The OECD defines such freely accessible educational materials as

> [...] digitized materials offered freely and openly for educators, students, and self-learners to use and reuse for teaching, learning, and research. OER includes learning content, software tools to develop, use, and distribute content, and implementation resources such as open licenses. (OECD 2007:29)

It is at this point that OER and MOOCs meet. A MOOC (Massive Open Online Course) is a form of networked learning that, in many ways, epitomizes the global network society in the area of education. MOOCs are forms of teaching and learning that encourage inter-organizational partnerships and user-generated content. MOOCs can be seen as an embodiment of a "connectivist" learning theory.[268] Connectivism is a theory of learning that goes beyond the psychological orientation of traditional learning theories such as behaviorism, cognitivism, and constructivism by claiming that knowing is not only a social activity but is distributed in networks. Connectivism shifts learning away from individuals and organizations and emphasizes the network as the primary subject and object of learning. Learning happens when information moves in all directions over networks rather than in the heads of individuals who are then required to prove knowledge acquisition by testing. MOOCs often use Open Educational Resources (OER), emphasizing user-generated content and participatory information production. MOOCs are educational platforms since they orchestrate resources, focus on interaction and sharing, and aim to improve the learning community.

[268] See Siemens (2005), Downes (2005, 2012), AlDahdouh et al. (2015) for discussions of connectivism. See also for an overview Wikipedia https://en.wikipedia.org/wiki/Connectivism.

There are different forms of MOOCs. Siemens and Downes base MOOCs on social learning, participation, user-generated content, and connectivism and find them in corporate training and normal certification programs of universities worldwide. Important consortia of institutions are Udacity, Coursera, edX, FUN, Iversity, and others. These MOOCs require educational management based on the open flow of information in networks.[269] Setting up and managing a MOOC requires that educational administration moves away from the traditional forms of information control that in the past have made it possible to design curricula, perform standard assessments, and carry out acknowledged certification.

Another example of how networking moves education beyond traditional boundaries is Personal Learning Environments (PLE). The idea of a PLE arose as an alternative to the centrally designed and top-down managed learning management systems (LMS), which often reflected institutional forms of organizing. In one form or another, LMS have become standard teaching tools in most educational institutions. Originally designed as a "virtual" classroom or virtual school, they carried the same centralism into the digital realm, top-down administrative procedures, teacher-centered pedagogical practices, and standard curricula and assessments typical of traditional schools.[270] However, the affordances of digital media and the tendency toward networking mitigate against conventional ways of organizing education. Consequently, new forms of delivering education that encouraged learner control, diversity of practices, and self-directed learning and communication were sought. The learners themselves and not the educational institutions should be making decisions about how and even what they learn, about processes, content, and information resources. The PLE also reflects the conviction that learning is a lifelong endeavor, extending beyond any institution, program of studies, or career qualification and including informal and formal learning, schooling, and on-the-job

---

269 See the contribution in the edutechwiki for a review of literature http://edutechwiki.unige.ch/en/E-learning_2.0.

270 For literature see the curated list of publications on PLE by Ilona Buchem https://ibuchem.wordpress.com/ple/.

training. However, lifelong learning is only practicable when learners themselves have something to say about the educational processes. Educational institutions can no longer assume that one-size-fits-all and that standardized instruction and assessment are the way to guarantee equality as well as quality in learning.

PLEs exemplify networking as opposed to systems in that they allow individuals access to a wide range of information resources, educational services, peer communities, aggregators, publication and distribution tools, such as blogs and wikis, and content management systems, such as e-portfolios. The PLE goes beyond any educational institution in that it constitutes a kind of hub for learner-centered, lifelong networked learning, both in regular schooling and in professional training, both formal and informal, both public and private. The PLE is multipurpose, heterogeneous, integrative, and adaptive. It is not an educational organization but a learning network no longer bound to a particular organization, institution, or degree program. It is learner-centered, open to many different applications, adaptable to other purposes and situations, and principally without limitations coming from external organizational constraints. The advent of PLEs has important consequences for educators and training administrators at all levels. Speaking for educational professionals and administrators, Leone writes, "The introduction of lifelong learning objectives and policies has poised us on the threshold of major change in education and society" (Leone 2010:30). Therefore, there is a "need for greater emphasis to be put upon flexibility, transferability, individualization, modularization, and mobility in education" (31). It has become a task for educational leaders to change from a system to a network.

Healthcare

If we look at healthcare, we can see how networks are coming into being that are not characterized by clear constitutive boundaries. Examples of this can be seen in the Quantified Self Movement and the various Platform or Health Ecosystem initiatives. One of the most interesting recent developments in networked healthcare, e-health, or connected health is the so-called "Quantified Self" (QS) movement.[271] Quantified self designates a broad range of new tracking technologies, devices, apps, services, practices, and values that all aim at "quantifying" health data: collecting, aggregating, evaluating, sharing, and commercializing personal health and wellness-related data. Other terms for quantified self are life-logging, body tracking, self-tracking, auto analytics, body hacking, and self-surveillance. A well-known slogan of the movement is "self-knowledge through numbers."[272] It is important to emphasize that this trend to track, quantify, and evaluate personal data is not limited to healthcare. Similar technologies and practices can be found in other areas. We have already discussed applications of this kind in education, for example, personal learning environments and learning analytics. The broader concept for personal data collection and analytics is "personal informatics." Personal informatics has become a general term describing many different kinds of digital self-monitoring concerning the body, health, and wellness, as well as other areas of life such as work, hobbies, and education. The official Personal Informatics website defines the term as follows:

> Personal informatics is a class of tools that help people collect personally relevant information for the purpose of self-reflection and self-monitoring. These tools help people gain self-knowledge about one's behaviors, habits, and thoughts. (www.personalinformatics.org)

Personal informatics includes not only the tracking of vital functions but also the monitoring of emotional and psychological states such as stress, happiness, sleep, and relaxation. Furthermore, there are many different technologies and services that can be classified as personal informatics, which relate to productivity at work, consumer behavior, monitoring learning processes, financial activities, hobbies, sports activities, etc. Within this broad understanding of self-monitoring, many different kinds of networks and organizations are involved. The data

[271] See https://en.wikipedia.org/wiki/Quantified_Self as well as the official website http://quantifiedself.com/.

[272] For an overview see Wikipedia https://en.wikipedia.org/wiki/Quantified_self.

that is gathered enters into and creates networks. For example, individuals are networked with providers, developers, employers, social and professional communities, educational institutions, businesses, government, healthcare organizations, and patient communities.

In these networks, individuals are not integrated into closed systems. They become administrators of their own information and make independent decisions about how this information is to be used. Personal informatics, therefore, offers many new opportunities, dangers, and challenges for decision-makers in organizations that are interested in using this data in different ways. Personal informatics is also closely linked to developments such as the Internet of Things, Ambient Assisted Living, smart buildings, smart neighborhoods, and smart cities. Making homes and cities smart is not simply building networked infrastructures but transforming closed systems into networked organizations. Smart government, smart education, smart healthcare, and smart organizations of all kinds relate to each other in ways that go beyond the mere adaptive behavior of systems.[273] All of these networks are connected to each other. No smart cities are without smart governments, smart education, smart businesses, etc. These classic domains of society can only become smart when they connect to each other and form new kinds of networks. Organizing the global network society necessarily crosses boundaries and links activities that were traditionally isolated in different domains, departments, jurisdictions, and social systems. From this perspective, the quantified self movement has many similarities to other societal changes. In business, education, and healthcare, networks cross institutional boundaries and branch out into many different organizational fields.[274]

Technically the quantified self movement relies on smartphones, smart watches, armbands, smart scales, wearables, smart clothes, and

---

[273] We see an example of this in the discussion of the Brazilian City of Curitiba.

[274] The digitalization of healthcare goes, of course, beyond QS and includes telehealth, telemedicine, medical e-learning, electronic patient records, hospital and clinical information systems, health data analytics, e-patient communities, shared decision making, and much more. For an overview see Belliger/Krieger (2014b),

implants, also referred to as "insideables," including injected or swallowed microchips and similar technologies. These devices are equipped with sensors that measure chemical, magnetic, mechanical, visual, acoustic, and thermal changes in the body or near environment. The data is registered and transmitted, usually via wireless or NFC technologies, to a smartphone or computer, which then evaluates it and sends it to the Internet, service providers, portals, community websites, etc. In this process, specially designed software or apps for mobile devices play an important role. Hundreds of thousands of health-related apps are available in the major app stores maintained by Google, Apple, and Microsoft.[275] These devices and apps continually monitor vital functions such as blood pressure, heart rate, calorie consumption, sleep quality, movement, glucose levels, stress, emotional states, and posture, as well as environmental influences such as UV radiation, air quality, etc. The evaluation, interpretation, and presentation of this data are done by many different services that the device or app producers usually offer. Some typical services are the aggregation of data, visualization of data, evaluation, comparison, sharing, recommendations and tips, health-related advice, e-consulting, referral to consultation by medical professionals and peers in patient communities, and personalized advertising by sellers of health-related products.

Self-tracking can become part of diagnosis, therapy, and rehabilitation processes. At this point, doctors, clinics, hospitals, laboratories, health insurance companies, and regulatory agencies also become involved in the network.[276] The same blurring of traditional boundaries can be observed in the areas of fitness, well-being, and prevention. Not only individuals are involved but also health insurers, public health programs, and prevention-related organizations, including employers concerned with having a healthy workforce. Quantified self is a new kind of network involving individuals across all demographic distinctions and many different kinds of organizations both within and without the traditional healthcare sector. Quantified self is changing the traditional

[275] Lutpon (2014a: 608) found over 100,000 health and medical apps mid 2014 in the Apple App Store and Google Play.

[276] See Lupton (2014c) for a discussion of medical apps and usage practices.

landscape of healthcare by blurring the boundary between therapy and prevention and empowering individuals to gather and evaluate their medical data, providing new data for medical and clinical research, enabling citizen science, and effectively disrupting the traditional, hierarchical healthcare system.

The dismantling of closed system boundaries is also clearly visible in the tendency to establish healthcare platforms. One can speak of a significant trend in healthcare: moving away from isolated, individual healthcare providers to networks of providers and patients linked together over platforms.[277] Benedict at al. (2018: 241) surveyed twenty-three eHealth platforms in Europe and concluded, "Within eHealth infrastructures we observe a predominant role of eHealth -platforms." These platforms bring together many different healthcare providers and offer customers various services in the areas of information, insurance, diagnosis, therapy, and prevention. Platforms usually consist of four different "roles." There is first and second the platform provider and or sponsor who set up and maintain the digital infrastructure. Third, the providers of products and services link together over the platform and establish a market for healthcare consumers. Fourth, there are the healthcare consumers, who are often prosumers in that they can sometimes actively participate in developing and testing products and services. The platform mediates transactions between all parties. Providers use the synergies of the cooperation via the platform to optimize business plans, avoid unfruitful competition, gather and use data, and maintain customer contact. The major advantages of eHealth platforms over the traditional healthcare system are efficiency, better quality of care, simple usability for all participants, accessibility for many different stakeholders, and empowerment of health services consumers. In addition, one could add that the greater volume and quality of the data generated by the platform enables the development of better services and new business models. Platforms, therefore, help reduce the time and effort required to access and manage health information, streamline administrative tasks, and thus reduce costs in many

[277] For a recent studies on eHealth platforms see Benedict et al. (2018); and Bernardo et al. (2022).

areas. Platforms are an integral part of the digital transformation of healthcare and a clear example of how networks are breaking out of the boundaries imposed by closed systems and even replacing systems in today's world.

4.3.3 Networks Cannot be Controlled Top-Down

Business

Systems are based on selection, relationing, and steering or control processes. No cybernetic system can exist without a "controller" of some kind. It is the reference to a controller that gives cybernetics its name. The *kybernetes* was the helmsman of a ship. Another word for the controller is governor. Turning to living systems, although there are some exceptions, the "central nervous system" is an excellent example of how living systems most efficiently steer their processes and operations. The same is true on the level of meaning. For centuries, if not thousands of years, centralized or hierarchical control has been a primary form of establishing and maintaining social order. It is hard to find any society that does not have a "political" function. The digital transformation and the advent of a global network society place this honored tradition in question. However, this is not a new form of anarchy. As the complexity of the world increases, hierarchies everywhere are becoming less efficient. In a complex, quickly changing environment, centralized decision-making brings many problems. These are 1) long and slow communication channels, 2) all the disadvantages of bureaucracy which hinder quick responses to problems and innovative solutions, 3) inflexibility regarding changing demands from within and outside the organization, 4) inefficient use of knowledge and information, 5) tendency to disempower people at lower levels of the organization, 6) lack of diversity and emphasis on standardization, 7) fostering of a culture of conformity, 8) mistrust and power struggles, 8) inability to manage remote work and dynamic teams which hinders collaboration, knowledge sharing, and collective decision-making.

Given these well-known problems facing traditional organizations, much effort has been made in the last decades to describe the challenges of "21st-century management."[278] A central topic that has emerged from this research is the position and function of management in networked organizations.[279] Although explicitly concerned with inter-organizational networks, the framework of network leadership and management developed by Saz-Carranza et al. (2008) can be applied to networks of all kinds. As often remarked (Powel 1990), networks appear as a form of order somewhere between the diversity and heterogeneity of markets and the unity of hierarchical, bureaucratic organizations. Viewing networks as a middle ground between markets and hierarchies with characteristics drawn from both may seem helpful. However, this view is still bound to the dualistic assumptions of agency and structure, or individual and organization typical of traditional organization theory. To say that networks are neither markets nor hierarchies does not necessarily imply that they are some mixture of the two, a middle ground between the opposing forces of agency and structure.

On the contrary, networks are neither markets, hierarchies, or a mixture of the two. They are something entirely different. This does not mean, however, that the task of managing networks successfully has nothing to do with forces of diversity and unity. Saz-Carranza et al. locate network management problems in the "inherent paradoxes implied by networks, in particular, the need to be simultaneously united and diverse" (2008: 292). This implies that "successful networks are simultaneously united and diverse" (297). Instead of finding a normative basis in contracts and property rights, as do markets or employer/employee relations, as in hierarchies, networks are based on complementarity, mutual adjustment, and reciprocity. They emerge and act within a "collaborative context" (Chrislip/Larson 1994) characterized by the specific "boundary-crossing nature of collaborations […], the lack of formal authority and hierarchy, and the blurriness of strategies" (Saz-Carranza et al. 2008:296). A collaborative context demands

---

[278] See the classic work of Peter Drucker (1999) and the extensive handbook by Wankel (2008).

[279] See Powel (1990), Jarrillo (1993), Huxham/Vangen (2000), Kickert et al. (1997), Koppenjan/Klijn (2004), and Saz-Carranza et al. (2008).

different leadership and management practices than those established in the industrial age.

> Leadership in collaborative contexts must be necessarily different, focusing largely on process, and has similarities to facilitative, transformative, and servant leadership – that is, to inspire commitment and action, to lead as peer problem solver, to build broad-based involvement, and to sustain hope and participation. (Saz-Carranza et al. 2008:296)

Citing Huxham and Vangen (2000), Saz-Carranza and colleagues define leadership broadly as "mechanisms that make things happen in a collaboration" (2008:296). Contrary to much research on leadership, which focuses on individuals conceived of as either leaders or followers, that is, on their skills, roles, competencies, talents, or authority, Saz-Carranza et al. propose focusing on "activities and actions that make things happen" (296). Therefore, they do not need to draw a clear boundary between management and leadership since bureaucratic routines and authoritative decisions no longer play important roles. Furthermore, the framework they propose does not focus on leadership and management *within* networks, that is, within network nodes or network partners, but instead describes leadership and management *of* the entire network. Leadership is a "collective phenomenon" (297) that distributes among all participants the task of "generating unity among the network members while preserving their diversity" (297). Enabling networks to be at once diverse and unified gives them an advantage over traditional forms of organization. Diversity fosters innovation, adaptability, flexibility, and resilience, while unity allows the various partners to coordinate action and to perform cooperatively and effectively.

> In sum, the unity/diversity paradox is inherent to networks, which must be diverse to have an added value with respect to hierarchies but united to allow for concerted action of any kind, unlike markets. Leadership is about avoiding diversity and unity to undermine each other by respectively generating disunity and similarity. (297)

Based on this understanding of the task that network management must accomplish, Saz-Carranza et al. develop a framework according to which network management practices fall into four categories: *facilitating, framing, activating,* and *mobilizing*.[280]

Facilitating includes all those management practices and activities that enable and support interaction, equality, motivation, and participation among network members. It is not merely a matter of exchange of information but of inclusive and participatory decision-making. Otherwise, certain members may feel disadvantaged, and the network would tend to disintegrate. *Facilitating* is a management practice derived from the network norms of connectivity, flow, and communication. These norms guide networking toward inclusivity while maintaining diversity, heterogeneity, open communication, and participatory action.

F*raming* is the second category of management practices Saz-Carranza and colleagues identify as peculiar to networked organizations. As the name suggests, framing includes all activities that enable cooperative action with regard to specific goals. This term strongly recalls Goffman's staging and Weick's sensemaking. It is also closely related to Latour's notion of "localizing" (Latour 1996:234) since framing sets up structures, rules, boundaries, etc., that focus, channel, partition, and reduce cooperative action. Framing consists of those management practices that operate not primarily for the network as a whole but for decentralized units, nodes, and partners, who can manage their unique contributions to the entire network in this way. Without framing, the diversity of the network would not be able to become operative since individual units would not have sufficient freedom and decision-making powers to organize their respective activities. Framing can be understood as guided by the network norms of participation and authenticity. Each partner can and should be responsible to the network in ways that they are themselves able to determine, at least partly. This demands management practices that make participation possible without unnecessarily limiting participants' authentic expression.

---

[280] Saz-Carranza et al. (2008: 297ff.).

Activating is the third form of management practice found in networks. It refers to all management practices that seek to integrate actors into the network. Activating is similar to what ANT calls translation and enrollment. The network becomes more powerful and effective to the extent that it can enroll new members who contribute new skills, competencies, ideas, and impulses. Just as translating and enrolling in ANT, activating in the network management framework of Saz-Carranza and colleagues is concerned with network goals or programs of action into which members can be integrated, and thus, it is also concerned with network identity, reputation, and legitimacy. The network norms of connectivity, participation, authenticity, and transparency play roles in successful activation since it is only when all participants in the network are accepted for their unique contributions and information is available to all that the network as a whole can become a collective actor capable of cooperative action concerning consensual goals.

Finally, *mobilization* refers to management practices that can bring resources and support into the network. The term "mobilization" in the Saz-Carranza framework means much the same as it does for ANT. For ANT, actor-networks "mobilize" as many human and non-human actors as possible to create stability, durability, social presence, and power. Translating and enrolling actors into a network becomes mobilizing when the goal is to generate network power. For Saz-Carranza and colleagues, "Mobilizing essentially builds network power […]" (298). Summarizing the four aspects of network management, they write:

> Sustaining the unity/diversity paradox generates power, which increases the network's effectiveness. The networks sustain the unity/diversity paradox by activating members, facilitating interaction, and framing the structure (procedures, rules, and values). In addition, networks mobilize support. When activating, the network selects and attracts members who share certain experiences, values, and principles but who are diverse regarding other organizational characteristics. Interaction and open decision making among the diverse members must be

> facilitated, and members united by framing common procedures, rules, and values. (2008:298-299)

This framework for understanding network management practices distinguishes itself from traditional management models because network management is not about decision-making but *facilitating, framing, activating,* and *mobilizing* decision-making, which no longer lies exclusively at the upper levels of the hierarchy. Decisions are distributed throughout the network and are made by network members. Management does not reduce complexity or uncertainty but establishes procedures for openness and inclusiveness. Furthermore, management does not control outcomes but facilitates processes. It does this through what Weick (1995) would undoubtedly call sensemaking. In words similar to how Weick describes sensemaking, Saz-Carranza et al. understand framing as what "ascribes, interprets, and makes meaning of the outside reality; it constructs the perceptions of actors regarding reality" (299).

The description of network management practices in facilitating, framing, activating, and mobilizing enables actors to "go through a naming and framing process, which is, basically, meaning making" (299). As we have shown in detail above, networking is a matter of translating, enrolling, and mobilizing actors into networks guided by programs of action and concerned with allowing all actors to have their own voice. In the global network society characterized by the affordances of new digital technologies, management practices of facilitating, framing, activating, and mobilizing can be derived from the norms of connectivity, flow, communication, participation, transparency, authenticity, and flexibility. It would be a mistake to view networking and networked forms of organizing as idealistic or ideologically motivated activism that ignores the realities of economics and market constraints. On the contrary, networked forms of organizing in all areas, whether public or private, profit or non-profit, can be said to have a competitive advantage over traditional organizations. Lavie (2008:324) notes that "firms can no longer be considered simply as independent entities competing for favorable market positions and protecting their core assets from imitation and appropriation." The global network society creates a situation where "firms have become interconnected in the sense that they engage in multiple simultaneous alliances" (324), contributing directly to economic success.

The fact that networks, as opposed to systems, are not steered effectively top-down is usually discussed in terms of flexibility. However, speaking of organizations as "flexible" often occurs within the framework offered by evolutionary theory and the theory of complex adaptive systems. Organizations are compared to biological organisms that can maintain their operations in the face of environmental changes by building up internal complexity. The more complex a system becomes, the more possible states it has, and thus, the more options it has to react to environmental changes. Since it became apparent that bureaucratic, hierarchical organization does not favor internal complexity, management theories first looked to systems theory for a model of how organizations can foster innovation. This model is adaptive behavior and learning, and a typical example of this theoretical development is Stafford Beer's Viable Systems Model of management.[281] Viable organizations tolerate internal complexity to adapt to a complex environment. Herbert Simon, for example, speaks of "nearly decomposable systems" (1996:193) that consist of relatively independent specialized elements. "The claim is that the potential for rapid evolution exists in any complex system that consists of a set of stable subsystems, each operating nearly independently of the detailed processes going on within the other subsystems" (193). It is well known that systems can only increase internal complexity to a certain extent or otherwise risk losing the central control that they need to pursue fixed goals and coordinate operations. The systems model has two drawbacks in this context. First, there is the issue of whether models of adaptive behavior do not lead to goals of sustainability, which are necessarily conservative and do not foster innovation. Second, we have noted above that the increasing complexity of systems tends to transform them into networks. From the point of view of network theory, there are no inherent limitations

---

[281] For an overview see Wikipedia https://en.wikipedia.org/wiki/Viable_system_theory.

on internal complexity. In networks, flexibility does not need to be achieved by special measures to increase internal complexity since networks do not arise as a complexity reduction in the first place. Flexibility is a characteristic of networks as such.

Networks are structurally opposed to hierarchies because every node in the network has a certain freedom to enter into connections to other nodes, bring new nodes into the network, and branch out into new directions with new goals. Of course, nodes in a network can be black-boxed into functional units with fixed input/output mechanisms. The prevalence of such black boxes everywhere is indisputable. This is why the systems model appears so plausible and convincing. Functional networks are needed and convenient for many reasons: efficiency, predictability, and control. Before the digital transformation, limitations on communication made it necessary to construct black boxes within organizations and sharp boundaries between organizations. Black boxes, however, can be opened, and any element of a functional unit can become a mediator again and enter into new and unforeseen associations. The affordances of digital technologies tend to open black boxes and deconstruct system boundaries. This characteristic of network structure makes complexity and flexibility into "normal" attributes of networks. Networks do not need to be made flexible by extraordinary measures. They are, by nature, flexible. What from the systems point of view is a complex and challenging task is for networks a matter of the normal dynamic of connectivity, flow, and participation. The more connectivity a network has, the more nodes it has and the more information flows through the nodes in unforeseeable ways, leading to more participation and unexpected developments, that is, to innovation and change.

Innovation from the systems theory point of view is a response to environmental changes that perturb the system but do not "inform" the system. Innovation for systems is always adaptation. Information construction is internal to the system. In a certain sense, it can be said that environmental changes are always threats to the system. If the system does not construct information adequately, it ceases to be viable.

System viability depends on being able to adapt quickly to environmental changes. Much of systems-based management theory is devoted to understanding how systems can become more adaptable, which is today called "resilience."[282] Risk management and innovation have become central topics in today's business world. Innovation, however, can be conceived as a synergy of organization and environment, a mutual and interdependent exchange.[283] In this view, both the organization and its environment mutually construct each other, which, we would argue, is another way of speaking about networks. Networks do not have an "environment" in the specific sense that complex adaptive systems do. Networks do not "adapt" to an environment in the same way that biological systems do. This has consequences not only for how innovation and organizational change are theorized but also for what has been called "organizational learning."[284]

Education

Trust-based management, decentralization, leveling out hierarchies, and involving stakeholders in decision-making are new forms of organizing that also have an important role in education. In administration, teaching, and learning, education is a cooperative effort in which everyone plays active roles. Therefore, it is not surprising that school administrators are also decentralizing, turning more to self-management, bottom-up management, and collaborative management to profit from experiences and resources available beyond the reach of one single school or group of instructors and teachers.[285] Outcomes of partnerships and networks are often uncertain and cannot be easily incorporated into budgets and quality control processes that can be centrally steered.

---

[282] A recent McKinsey survey underlines the importance that the term as gained in the business world. See https://www.mckinsey.com/capabilities/risk-and-resilience/our-insights/from-risk-management-to-strategic-resilience

[283] "In the 21st century, organizations should not solely respond to preordained environmental conditions, but should instead influence and actually create their environment by innovation" (Kollman/Stöckmann 2008:12).

[284] For an overview of research on organizational learning and knowledge management, see Easterby-Smith/Lyles (2011).

[285] See for example the discussion on school partnerships in Muijs et al. (2011).

Accountability, as crucial as it is, above all in the public sector where much education is located, must be balanced against innovation and risk-taking for long-term improvements in teaching, development of learning materials, learning success, assessment, and curricula reform. These processes show that partnerships and networking can make resources and infrastructures accessible so that information is distributed to all participants and instruments and procedures are set up to transfer knowledge throughout and beyond the network.

One of the most important trends in the use of digital information today in all areas is so-called "big data." Big data refers to the availability of very large data sets (volume) from diverse sources and in many different formats (variety), which can be "analyzed" quickly, if not in real-time (velocity), by special analytic software.[286] Analytics aims to discover trends and correlations in large datasets. Only when data becomes big in volume, variety, and velocity, do correlations and patterns that can otherwise not be found become apparent. This information can be useful for descriptive, predictive, preventive, and prescriptive analytics. These new possibilities of using data affect two significant areas in educational management. These are networked learning and the digitalization of educational administration. One speaks of "learning analytics" in both areas to describe all aspects of big data in education. Jones (2015) speaks of learner analytics to identify "those aspects of learning analytics that focus on the learner or student as opposed to the institutional or business aspects of analytics" (114). We have spoken above about how personal learning environments and MOOCs extend teaching and learning beyond the borders of traditional institutions.

Learning analytics can illustrate how top-down management in education is becoming inefficient and questionable. Learning analytics applications aim to predict and identify which students need support and allow instructors to intervene to optimize learning outcomes.[287] It has

[286] See https://en.wikipedia.org/wiki/Big_data for an overview.

[287] See the discussion and examples of applications in EduFutures.net http://edfutures.net/Learning_Analytics.

become possible to track not only online activities within a traditional LMS but also to monitor social media usage, search activities on the Internet, emails, and physical activities such as using the library, sports, or entertainment facilities on and off campus. This kind of tracking of student activities, as with other big data applications outside the educational context, creates an entirely new situation. The learner becomes a "glass human being" who is completely transparent to those with access to the data and analytic applications. This results in a situation in which administrators are forced to include all stakeholders in decision-making. Jones (2015) cites the Director of Teaching and Learning at the Open University UK, who voiced his concern as follows:

> With these possibilities come dangers that the data could be used in ways undesirable to students. These include invading their privacy, exploiting them commercially by selling their data to third parties or targeted marketing of further educational products. (120)

In the area of enterprise management of educational institutions, the same problems occur. Analytics allows for a kind of educational "business intelligence" in which not only students are being monitored but teaching staff, non-teaching employees, and other operations as well. Big data is used for ERP systems to monitor not only libraries, laboratories, and materials suppliers for teaching and learning, but also customer relations, housing, parking, catering, athletics, etc. There are no technical barriers to data use in all areas, leading to the problem that business imperatives could be applied to teaching and learning. As Jones points out,

> If a broad administrative view of analytics is taken, then analytics can be just another management and administrative tool that will shift the balance in institutions away from academic and pedagogic concerns towards market and business concerns with measurable performance and value for money. Measurements of quality, progression and dropout when seen in terms of students as units of resource are quite different from

> measures of quality understood in terms of human development and citizenship. (Jones 2015:115)

The challenge to education management coming from big data and learning analytics is to break the patterns of traditional business management and adopt new *symmetrical transparency* regarding the uses of data. In the long run, analytics cannot succeed as a top-down administrative process. All stakeholders should be involved in decisions about what data will be collected and analyzed for what purpose, and all stakeholders should have equal access to the results of analytics. This implies that these results be made understandable and useable by all stakeholders, including learners. The unavoidable transparency of all users, whether students or staff, should be compensated by symmetrical transparency on the side of administration, decision-makers, and institutional leaders. Transforming top-down transparency into bottom-up transparency can only be successful based on open communication and trust. Jones (116) discusses the framework developed by De Laat in which users are themselves responsible for "tagging problems that are part of their learning," "adding people they collaborate with," and "linking themselves with problems others have already described." Networking transforms education into a self-organizing process and forces a revision of traditional hierarchical management practices. Within a framework such as this, learning analytics is actively implemented by users so that more and better information is collected, leading to more reliable and more valuable results. Instead of being monitored behind their backs, as it were, all participants actively contribute to data aggregation and consent to data use.

Healthcare

It is well known that healthcare is one of society's most heavily regulated sectors. It is also a well-established tradition that doctors are "gods in white" who control all health-related information about patients and often subscribe to a paternalistic model in which "the doctor knows best," and only the doctor decides what information to share with patients. There is hardly any other subsystem of society that is more characterized by hierarchy than the healthcare system. We have already discussed how the digital transformation and the shift from systems to networks are changing healthcare. E-Patients are now everywhere acknowledged and often welcomed. Medical information is now available to patients as never before. Patients are networked with other patients and medical professionals via platforms and communities. The quantified self movement, which is more active in prevention than primary care, also shows how networking and digital technologies empower people to take control of their health and well-being instead of relying on medical professionals. These developments challenge, disrupt, and transform the traditional hierarchical healthcare organization. We have discussed many of these developments above and will limit our discussion to only a few examples.

As an example of a bottom-up healthcare organization, the Dutch Buurtzorg model for patient-centered care is often cited and discussed as an exemplary solution to home-based healthcare.[288] The Dutch word *Buurtzorg* means "neighborhood care" and focuses on nursing and community health, not hospital treatment. It was founded in 2007 by a Dutch nurse, Jos de Blok, and has since become internationally recognized for its innovative and practical approach to delivering home healthcare services. The major characteristics of the Buurtzog model are self-managing teams of nurses consisting of ten to twelve, who are responsible for providing a wide range of healthcare services to patients that are located within a particular area. The teams come to know the patients personally and care not only for their physical health but also for the emotional, psychological, and social well-being of their patients. The emphasis is on the overall quality of life and not merely on whether someone can be classified as either sick or well. This concept of encompassing care assists patients in becoming more self-sufficient and responsible regarding their health. For this reason, much effort is devoted to personalized care, that is, working closely with patients and supporting family members to develop approaches to care that fit the personal needs of patients and their families.

---

[288] See https://www.buurtzorg.com/about-us/buurtzorgmodel/.

With regard to administration, Buurtzorg strives to reduce costs and bureaucracy. The aim is to allow nurses to spend more time with patients and less on paperwork, thus enhancing the quality of care. Furthermore, the teams emphasize preventive measures and health education to help patients maintain their health and well-being, reducing the need for more intensive interventions in the future. The nurses work collaboratively with other healthcare professionals and social workers and use available community resources to ensure comprehensive and coordinated care for their patients. Another important characteristic of this bottom-up approach to healthcare is increased flexibility in scheduling and care delivery. Nurses can decide how they manage their schedules to accommodate patients' needs and provide care when it's most convenient and effective.

Illustrating the digital transformation's values, Buurtzorg utilizes technologies such as a custom-built electronic health record system, digital support for communication and coordination, and electronic documentation. These measures empower the nurses to make decisions about patient care, how to manage the teams, and how to set up operational processes of their work. From this empowerment and autonomy, which one does not have in hierarchical organizations, the nurses share a sense of ownership and engagement lacking in traditional top-down organizations. It has been shown that the model effectively improves patient outcomes, reduces hospital admissions, and enhances patient and staff satisfaction. It is based on patient-centered care, decentralized decision-making, and community engagement.

Similar to Buurtzorg, many Community Health Worker programs have been initiated all over the world. These programs train and deploy individuals from within local communities to provide needed healthcare services and health education to people who would otherwise not have easy access to these services. Some examples, which are only mentioned here, are: the BRAC Manoshi in Bangladesh employs female community health workers to provide maternal and child healthcare services, family planning education, and health counseling to women and families in urban slums; the Treatment and Care for Everyone from South

Africa which is a community-based HIV prevention and care program; the Accredited Social Health Activists (ASHAs) program which is part of India's National Rural Health Mission; the Promotoras de Salud in the USA and Latin America, which employs community health promoters to improve healthcare in Latin American countries and Hispanic communities in the USA; the Mitanin Program in Chhattisgarh, India, which trains women from tribal and rural areas as community health workers; the Community Health Extension Workers in Nigeria who are trained to provide healthcare services such as maternal and child health, immunizations, family planning, and health education in communities. There are many more examples of this kind that could be cited here. The point is that there are clear tendencies to support bottom-up healthcare initiatives and thus shift decision-making to lower levels of the hierarchy.

Summary

We have attempted to give some concrete examples of the shift from systems to networks in contemporary society. It is admitted that the above discussion is neither exhaustive nor representative of how networking is changing society. And, of course, there can be no question that the examples we have described in business, education, and health-care in any way "prove" the thesis that we are currently going through a paradigm change in our understanding of the world and how we should organize our lives. Nonetheless, we hope to have convincingly shown that something is indeed changing, that what can plausibly be described as networked forms of social order are being adopted, and that they represent a significant departure from what could be called modern Western industrial society. This is the society that Luhmann's theory of social systems claimed to have theoretically modeled. Modern society, according to Luhmann, is a functionally differentiated society. Despite Luhmann's posthumanism and acceptance of globalization, systems theory is a theory, at least on the level of the social sciences, of Western modernity. Differently, Latour's actor-network theory is also a theory, or rather, an ethnology of the "Moderns." Latour, however, is convinced, and this is what he attempts to show in his empirical studies, that modernity has come to an end and that the task of thinking, of theory, of science has now become the task of assembling the collective.

We have called this new collective the global network society and have proposed in this book to have gone beyond Latour in formulating the values that guide networking. Unlike Latour, who does not address the question of the digital transformation, we have derived the values of the global network society from the affordances of digital technologies. After all, if society, as Latour says, is technology, it makes a difference for society what that technology is. The digital transformation is not only a technological change but a transformation of society. When society changes, values change. How are these changes to be theoretically modeled? What concepts, foundational assumptions, and future perspectives are needed to move constructively into the global future, wherein digital technologies, among them AI, will play a major role in how we live our lives, understand ourselves, and construct social order? These were the questions that motivated this book, and which have guided our discussion. We hope they are the right questions and that we have offered answers that at least are a small contribution to answering them.

Literature

Adams, K. M., Hester, P. T., Bradley, J. M. (2013): A Historical Perspective of Systems Theory. In: Proceedings of the 2013 Industrial and Systems Engineering Research Conference, A. Krishnamurthy and W.K.V. Chan, eds.

Ainamo. A. (2008): "Global Projects as New Organizational Form," in: Wankel 2008, Vol. 1, 482-489.

AlDahdouh, A. A., Osorio, A. J., Caires, S. (2015): "Understanding Knowledge Network Learning and Connectivism." in: International Journal of Instruction Technology and Distance Learning, Oct. Vol. 12. No. 10, 3-46.

Andersen, B.P., Miller, M., Vervaeke, J. (2022): Predictive processing and relevance realization: exploring convergent solutions to the frame problem. Phenom Cogn Sci.

Apel, K.-O. (19760): Transformation der Philosophie, 1. vol., Frankfurt: Suhrkamp.

Apel, K.-O. (1976): Transformation der Philosophie, 2. vol., Frankfurt: Suhrkamp.

Apel, K.-O. (1988): Understanding and Explanation: A Transcendental-Pragmatic Perspective, Boston: MIT Press.

Arsenault, A. (2011): "Networks: The Technological and the Social," in: Dealnty, G./Turner, S. P. (eds) Routledge International Handbook of Contemporary Social and Political Theory, 259-269. UK: Routledge.

Asbhy W.R. (1956): An Introduction to Cybernetics. London: Chapman & Hall.

Autio, T., Ropo, E. (2005): "Networking as a Strategy for Restructuring Basic and Further Teacher Education," in: Wiel, V., O'Hair, M. J. (eds) Network Learning for Educational Change. UK: Open University Press.

Baecker, D. (ed) (1993a): Problems of Form, Frankfurt: Suhrkamp.

Baecker, D. (1993b): Calculus of Form, Frankfurt: Suhrkamp.

Barabasi, L.-A. (2002): Linked. The New Science of Networks. Perseus Publishing: Cambridge MA.

Barabasi, L.-A. (2016): Network Science. Cambridge University Press.

Barney, J. (1991): "Firm Resources and Sustained Competitive Advantage,"

Journal of Management, 17(1), 99-117.

Bateson, G. (1972): Steps to an Ecology of Mind, New York: Ballentine.

Bateson, G. (1979): Mind and Nature, A Necessary Unity. New York: Dutton.

Beer, S. (1959): Cybernetics and Management, English University Press.

Belliger, A. (2008): "Wissensmanagement – ein Instrument auf dem Weg zur Lernenden Organisation," in: Schweizerische Zeitschrift für Heilpädgogik 9.

Belliger, A. (2010a): "Enterprise Social Networking. Kollaboration im Unternehmen 2.0: Organisation oder Netzwerk?," in: Wissensmanagement. Zeitschrift für Führungskräfte, 5/2010.

Belliger, A. (2010b): "Human Resource Management in Zeiten von Facebook, Twitter, iPhone & Co. Von Unternehmen als Netzwerken, Kontextsteuerung und Grenzmanagement," in: Risikofaktor Personal. Wie viel Kontrolle ist nötig. Management von Unsicherheit. Zeitschrift HR Today.

Belliger, A. (2010c): "eLearning und Wissensmanagement: Chancen und Herausforderungen von elektronischen Lern-Hilfsmitteln für Nutzer und Anbieter; wie eLearning effizient eingesetzt und genutzt werden kann," in: Organisator. Das Magazin für KMU, 7-8/2010.

Belliger, A. (2010d): "Die Akteur-Netzwerk-Theorie – oder: Wie sozial sind soziale Netzwerke?" in: Wissensmanagement. Zeitschrift für Führungskräfte, 6/2010

Belliger, A. (2011a): "Lernen und Lehren mit Technologien aus der Sicht der Akteur-Netzwerk-Theorie," in: Lehrbuch für Lernen und Lehren mit Technologien (L3T), TU Graz.

Belliger, A. (2011b): Social Media und Wissensnetzwerke in der Personalentwicklung, in: Jahrbuch Personalentwicklung 2012, Luchterhand.

Belliger, A. (2016): Digitale Transformation im Gesundheitswesen, in: Tagungsband zum 5. Nationalen Biobanken-Symposium der Technologie– und Methodenplattform für die vernetzte medizinische Forschung TMF, Berlin, Bd. 14 Schriftenreihe der TMF.

Belliger, A. (2017): Connected Health – Connected Architecture, in: Tagungsband 6. Symposium Health Care der Zukunft – Livability of Health, Technical University Berlin.

Belliger, A. (2013a): "Werte und Normen vernetzter Organisationen oder Digitale Geschäftsprozesse sind nur der Anfang," in: Wirtschaftsmagazin 1/2013.

Belliger, A. (2013b): "Social Media und Wissensnetzwerke in der Personalentwicklung, in: Handbuch eLearning," Wolters Kluwer Verlag (gemeinsam mit D. Krieger), 6.25, 1-10.

Belliger, A. (2014): "Vernetzte Gesundheit," in: Belliger/Krieger (eds) Gesundheit 2.0. Das ePatienten-Handbuch. Bielfelfd: Transcipt, 97-135.

Belliger, A., Krieger, D. J. (eds) (2006): ANThology: Ein einführendes Handbuch zur Akteur-Netzwerk-Theorie, transcript Verlag: Bielefeld,

Belliger, A., Krieger, D. J. (2008): "Netzwerk," in: Lexikon Soziologie und Sozialtheorie. Hundert Grundbegriffe, Stuttgart: Reclam.

Belliger, A., Krieger, D. J. (2009a): "ANT: Nichts Neues? Nichts Nützliches? Eine Replik zu Gesa Lindemann," in: Zeitschrift für Kulturwissenschaften.

Belliger, A., Krieger, D. J. (2009b): „Virtuelle Grenzen und Räume in Organisationen," in: G. Thomann, G./Bucher, Th/Hagmann, Th./Kuhn, R. (eds) Grenzmanagement, Resonanz Bd. 1,71-84, Bern: hep-verlag.

Belliger, A., Krieger, D. J. (eds) (2014): Gesundheit 2.0. Das ePatienten Handbuch, transcript Verlag: Bielefeld.

Belliger, A., Krieger, D. J. (2014a): Interpreting Networks – Hermeneutics, Actor-Network Theory & New Media, transcript Verlag: Bielefeld.

Belliger, A., Krieger, D. J. (2014b): "Netzwerke der Dinge (aus der Sicht der Akteur-Netzwerk-Theorie)," in: Samida, M. K. H./Eggert, H. P/ Hahn, S. (eds) Handbuch Materielle Kultur. Bedeutungen, Konzepte, Disziplinen, J.B. Metzler.

Belliger, A., Krieger, D. J. (2015): Die Selbstquantifizierung als Ritual virtualisierter Körperlichkeit, in: Körper und Ritual. Sozial- und kulturwissenschaftliche Zugänge und Analysen, Gugutzer, R., Staack, M. (eds), Springer: Wiesbaden, 389–404.

Belliger, A., Krieger, D. J. (2016a): From Quantified Self to Qualified Self – A Fictional Dialogue at the Mall, in: Quantified Selves Statistic Bodies, Digital Culture & Society Journal. P. Abend, M. Fuchs (eds), transcript Verlag: Bielefeld, S. 25–40.

Belliger, A., Krieger, D. J. (2016b): Organizing Networks – An Actor-Network Theory of Organizations, transcript Verlag: Bielefeld.

Belliger, A., Krieger, D. J. (2016c): The End of Media – Reconstructing Media Studies on the Basis of Actor-Network Theory, in: Applying the Actor-Network Theory in Media Studies, B. Ochsner, M. Spöhrer (Eds.), IGI Global, Hershey, PA, S. 20–37 http://www.igi-global.com/book/applying-actor-network-theory-media/147546

Belliger, A., Krieger, D. J. (2018): Network Publicy Governance – On Privacy and the Informational Self, transcript Verlag: Bielefeld.

Belliger, A., Krieger, D. J. (2021): Hacking Digital Ethics, Anthem Press. UK.

Belliger, A., Krieger, D. J. (2022): New Directions in Digital Ethics, in Casas-Roma, J., Conesa, J., Caballé, S. (eds) Technology Users and Uses – Ethics and Human Interaction Through Technology and AI, Ethics International Press: Cambridge UK.

Benedict M, Herrmann H, Esswein W. (2018): eHealth-Platforms – The Case of Europe. Stud Health Technol Inform. 247:241-245.

Berger, P. L. and T. Luckmann (1966), The Social Construction of Reality: A Treatise in the Sociology of Knowledge, Garden City, NY: Anchor Books.

Bernardo, J.; Apóstolo, J.; Loureiro, R.; Santana, E.; Yaylagul, N.K.; Dantas, C.; Ventura, F.; Duque, F.M.; Jøranson, N.; Zechner, M.; et al. (2022): eHealth Platforms to Promote Autonomous Life and Active Aging.

Bertalanffy, L. v. (1968): General Systems Theory, New York: George Braziller.

Bijker, W. E. (1995): Of Bicycles, Bakelites and Bulbs. Toward a Theory of Sociotechnical Change. Cambridge, MA: MIT Press.

Bijker, W. E./Hughes, Th. P./Pinch, T. (eds) (1987): The Social Construction of Technological Systems: New Directions in the Sociology and History of Technology. Cambridge, MA: MIT Press.

Boulding, K. E. (1956): General Systems Theory – Applications for Organization and Management. Management science, 2(3), 197-208.

Callon, M. (1987): Society in the Making: The Study of Technology as a Tool For Sociological Analysis, in Bijker, W./Hughes, T./Pinch, T. (eds) The Social Construction of Technological Systems. Cambridge: MIT Press.

Callon, M. (1991): Techno-Economic Networks and Irreversibility, in Law, J. (ed) A Sociology of Monsters: Essays on Power, Technology and Domination. London: Routledge.

Callon, M. (1986): "Elements of a sociology of translation: Domestication of the Scallops and the Fishermen of St Brieuc Bay". In John Law (Ed.), Power, Action and Belief: A New Sociology of Knowledge? London, Routledge: 196-233.

Callon, M., & Law, J. (1982). On interests and their transformation: enrolment and counter-enrolment. Social Studies of Science, 12(4), 615-625.

Callon, M., Law, J., & Rip, A. (1986). Mapping the dynamics of science and technology. Sociology of Science, 16(3), 397-441.

Callon, M., & Latour, B. (1987). Unscrewing the big Leviathan: How actors macro-structure reality and how sociologists help them to do so. In K. Knorr-Cetina & A. Cicourel (Eds.), Advances in social theory and methodology: Toward an integration of micro-and macro-sociologies (pp. 277-303). Routledge.

Callon, M. (1988). The role of lay people in the production and dissemination of scientific knowledge. Science, Technology, & Human Values, 13(1), 3-28.

Callon, M., & Law, J. (1995). Agency and the hybrid collectif. South Atlantic Quarterly, 94(2), 481-507.

Castells, M. (1996): The Rise of the Network Society. The Information Age: Economy, Society and Culture Vol. I, 2nd ed. 2000. Oxford: Blackwell.

Castells, M (1997a): The Power of Identity. The Information Age: Economy, Society and Culture Vol. II, 2nd ed. 2004. Oxford: Blackwell.

Castells, M. (1997b): An Introduction to the Information Age, in City 7, 6-16.

Castells, M (1998): End of Millennium. The Information Age: Economy, Society and Culture Vol. III. 2nd ed. 2000. Oxford: Blackwell.

Castells, M. (2000): Materials for an Exploratory Theory of the Network Society, in British Journal of Sociology Vol. No. 51 Issue No. 1 (January/ March 2000) pp. 5–24.

Castells, M. (2001): The Internet Galaxy. Oxford: Oxford University Press.

Castells, M. (2004): Informationalism, Networks, and the Network Society: A Theoretical Blueprint, in Castells, M. (ed) (2004): The Network Society. A Cross-cultural Perspective. UK: Edward Elgar, 3-45.

Castells, M. (2005): The Network Society: From Knowledge to Policy, in Castells, M./Cardoso, G. (eds) The Network Society from Knowledge to Policy. Washington, DC: Center for Transatlantic Relations.

Castells, M. (2009): Communication Power. New York: Oxford University Press.

Castells, M. (2011): A Network Theory of Power, in International Journal of Communication 5, 773-787.

Collis, B. (2005): E-Learning and the Transformation of Education for a Knowledge Economy, in Castells, M., Cordosa G. 2005, 215-223.

Cronin, L., Walker, S. (2023): Assembly Theory Explains and Quantifies the Emergence of Selection and Evolution. arXiv:2206.02279v3 [physics. bio-ph].

Czarniawska, B., Hernes, T. (eds) (2005): Actor-Network Theory and Organizing. Liber

Czarniawska, B. (1998): Narrative approach in organization studies. UK: Sage.

Czarniawska, B. (2004): Narratives in Social Science Research. UK: Sage.

Czarniawska, B. (2008): A Theory of Organizing. London. UK: Edward Elgar.

Deely, J. (1990): Basics of Semiotics, Bloomington: Indiana University Press.

Deplazes A, Huppenbauer M. (2009): Synthetic organisms and living machines: Positioning the products of synthetic biology at the borderline between living and non-living matter. Syst Synth Biol. 2009 Dec;3(1-4):55-63.

d'Espagnat, B. (2006) On Physics and Philosophy. Princeton University Press. Princeton & Oxford.

DiMaggio, P. J./Powell, W. (eds) (1991): The New Institutionalism in Organization Analysis. Chicago: University of Chicago Press.

Drucker, P. F. (1999): Management Challenges for the 21st Century. New York: Haprer.

Ebeling, W. (1989): Chaos – Order – Information. Selbstorganisation in Natur und Technik, Leipzig: Urania Verlag.

Ehlers, U. D. (2011): From open educational resources to open educational practices, in Elearning Papers, 23, 1–8. http://openeducationeuropa.eu/en/paper/open-educationchanging-educational-practices.

Fields, C., Friston, K., Glazebrook, JF., Levin, M. (2022): A free energy principle for generic quantum systems 1 Sep 2022, in Progress in Biophysics and Molecular Biology173:36-59.

Flechtner, H.-J. (1970): Grundbegriffe der Kybernetik. Stuttgart: Wissenschaftliche Verlagsgesellschaft.

Floridi, L, (2005): The Ontological Interpretation of Informational Privacy, in Ethics and Informational Technology, vol. 7, no. 4, pp. 185-220.

Floridi, L. (2014). The Fourth Revolution: How the Infosphere is Reshaping Human Reality, Oxford: Oxford University Press.

Floridi, L. (2016): On Human Dignity as a Foundation for the Right to Privacy, in: Philos. Technol. 29, S. 307-312.

Foerster, H. von 1992: Einführung in den Konstruktivismus. Munich: Piper.

Foerster, H. von 1993: Wissen und Gewissen, Frankfurt: Suhrkamp.

Friston, K. (2010): The free-energy principle: a unified brain theory?. Nat Rev Neurosci 11, 127–138. https://doi.org/10.1038/nrn2787

Friston K, Frith C. D. (2015): Active inference, communication and hermeneutics. Cortex. 2015 Jul; 68:129-43. doi: 10.1016/j.cortex.2015.03.025.

Gadamer, H.-G. 1989: Truth and Method, 2nd edn. London: Sheed and Ward.

Gamble, C., Poor, M. (eds) (2005): The Hominid Individual in Context. Archeological Investigations of Lowe and Middle Paleolithic Landscape, Locales and Artefacts.

Gerok, W. (1990): Ordnung und Chaos in der unbelebten und belebten Natur, Stuttgart: Wissenschaftliche Buchgesellschaft.

Gibson, J. (1977): The Theory of Affordances, in Shaw, R./Bransford, J. (eds) Perceiving, Acting, and Knowing: Toward an Ecological Psychology. New York: Wiley.

Giddens, A. (1984): The constitution of society: Outline of the theory of structuration. Cambridge: Polity Press.

Glasersfeld, E. von (1987a): Knowledge, Language and Reality. Works on radical constructivism, Braunschweig: Vieweg.

Glasersfeld, E. von (1987b): Siegener Gespräche über Radikalen Konstruktivismus, in: Schmidt 1987.

Glasersfeld, E. von (1992): Construction of Reality and the Concept of Objectivity, in: Introduction to Constructivism, Munich: Piper.

Gumbrecht, H. U., Pfeiffer, K. L. (eds) (1994): Materialities of Communication, Stanford U. Press: Stanford CA.

Habermas, J. (1971): Theory of Society or Social Technology?, in Habermas/ Luhmann 1971.

Habermas, J. (1976): Legitimation Crisis, T. McCarthy (trans), Cambridge: Politiy Press.

Habermas, J. (1981): Theory of Communicative Action. 2 Volumes. McCarthy, Th. (trans). Boston, MA: Beacon Press

Habermas, J. (1989): The Structural Transformation of the Public Sphere: An Inquiry into a Category of Bourgeois Society, Cambridge: Polity Press.

Habermas, J. (1984): Preliminary Studies and Additions to the Theory of Communicative Action, Frankfurt: Suhrkamp.

Habermas, J. (1985): The Philosophical Discourse of Modernity, Frankfurt: Suhrkamp.

Habermas, J., Luhmann, N. (eds) (1971): Theory of Society or Social Technology. What does systems research achieve? Frankfurt: Suhrkamp.

Haferkamp, H., Schmid, M. (eds) (1987): Meaning, Communication and Social Differentiation. Beiträge zu Luhmanns Theorie sozialer Systeme, Frankfurt: Suhrkamp.

Haken, H., Haken-Krell, M. (1989): Emergence of Biological Information and Order, Darmstadt: Wissenschaftliche Buchgesellschaft.

Haken, H., Wunderlin, A. (1991): The Self-Structuring of Matter. Synercgetics in the Inanimate World, Braunschweig: Vieweg.

Heidegger, M. [1927] (1962): Being and Time. Macquarie, J. (trans). Oxford: Blackwell.

Heidegger, M. (1977): The Question Concerning Technology, in Basic Writings Krell, D. F. (ed), New York: Harper & Row, 287.

Hidalgo, C. A. (2016): Disconnected, Fragmented, or United? A Trans-Disciplanary Review of Network Science. Appl Netw Sci 1, 6.

Hoffman, D. (2019): The Case Against Reality: How Evolution Hid the Truth from Our Eyes, Allen Lane.

Hutchins, E. (1995): Cognition in the Wild. Cambridge, MA: MIT Press.

Jantsch, E. (1980): The Self-Organizing Universe. Pergamon Press Oxford/ New York.

Jantsch, E. (1987): Epistemological Aspects of the Self-Organization of Natural Systems, in Schmidt 1987.

Jarrillo, J. C. (1993): Strategic Networks: The Borderless Organisation. Oxford: Butterworth-Heinemann.

Jones, Ch. (2015): Networked Learning: An Educational Paradigm for the

Age of Digital Networks. Switzerland: Springer.

Kaduk, S., Osmetz, D., Wüthrich, H. A., Hammer, D. (2015): Müsterbrecher. Die Kunst, das Spiel zu drehen. Hamburg: Murmann.

Katz, J., Gartner, W. B. (1988): Properties of Emerging Organizations, in Academy of Management Review, 13(3), 429-441.

Katz, J. A., Shepherd, D. A. (2004): Corporate Entrepreneurship. New York: Elsevier.

Katz M. L., Shapiro C. (1985): Network Externalities, Competition, and Compatibility, in American Economic Review 75: 424-440.

Katz, M. L., Shapiro, C. (1994): Systems Competition and Network Effects, in Journal of Economic Perspectives 8: 93-115.

Kalpokas, I. (2919): Algorithmic Governance. Politics and Law, in the Post-Human Era. Palgrave Macmillan: London.

Kast, F. E., Rosenzweig, J. E. (1972): General systems theory: Applications for organization and management, in Academy of management journal, 15(4), 447-465.

Kiani, N., Gomez-Cabrero, D., Bianconi, G. (eds) (2021): Networks of Networks, in Biology: Concepts, Tools and Applications. Cambridge: Cambridge University Press.

Kickert, W., Klijn, E.-H., Koopenjan, J. (1997): Managing Complex Networks. London: Sage.

Klaus, H., Liebscher, G. (1979): Wörterbuch der Kybernetik. Fischer Verlag.

Kneer, G. (1992): Conservation and Reflection. On the critical reformulation of social rationality, in Krawietz / Welker 1992.

Kneer, G., Nassehi, A. (eds) (1993): Niklas Luhmanns Theorie sozialer Systeme. Eine Einführung, München: Fink.

Kratky, K. W., Wallner, F. (eds) (1990): Grundprinzipien der Selstorganisation, Darmstadt: Wissenschaftliche Buchgesellschaft.

Krawietz, W., Welker, M. (eds) (1992): Kritik der Theorie sozialer Systeme. Auseinandersetzungen mit Luhmanns Hauptwerk, Frankfurt: Suhrkamp.

Krieger, D. J., Belliger, A. (2014): Interpreting Networks – Hermeneutics, Actor-Networ Theory and New Media. Bielefeld: transcript Verlag.

Krohn, W., Küppers, G. (eds) (1989): Die Selbstorganisation der Wissenschaft. Frankfurt: Suhrkamp.

Krohn, W., Küppers, G. (1992): Emergenz. Die Entstehung von Ordnung, Organisation und Bedeutung, Frankfurt: Suhrkamp.

Latour, B. (1987): Science in Action: How to Follow Scientists and Engineers Through Society. Cambridge, MA: Harvard University Press.

Latour, B. (1992): Where Are the Missing Masses? The Sociology of a Few Mundane Artifacts, in Bijker, W. E., Law, J. (eds) Shaping Technology/ Building Society: Studies in Sociotechnical Change. Cambridge, MA: MIT Press.

Latour, B. (1993a): We Have Never Been Modern. Cambridge, MA: Harvard University Press.

Latour, B. (1993b): Pasteurization of France. Cambridge, MA: Harvard U. Press.

Latour, B. (1994): On Technical Mediation, in Common Knowledge Vol.3, n2: 29-64.

Latour, B. (1996a): On Interobjectivity, Mind, Culture, and Activity. Vol. 3, No. 4: 228-245.

Latour, B. (1996b): Review of Ed Hutchins' Cognition in the Wild, in Mind, Culture, and Activity. Vol. 3, No. 1: 54-63.

Latour, B. (1999): Pandora's Hope: Essays on the Reality of Science Studies, Cambridge MA: Harvard University Press.

Latour, B. (2002): War of the Worlds: What about Pace? Chicago: Prickly Paradigm Press.

Latour, B. (2004a): Why Has Critique Run out of Steam? From Matters of Fact ot Matters of Concern, in Critical Inquiry 30 (Winter): 225-248. Chicago: University of Chicago Press.

Latour, B. (2004b). Politics of Nature. How to Bring the Sciences into Democracy. Cambridge MA: Harvard University Press.

Latour, B. (2005). Reassembling the social: An introduction to actor-network theory. Oxford University Press.

Latour, B. (2008): A Cautious Prometheus? A Few Steps Toward a Philosophy of Design (with Special Attention to Peter Sloterdijk), in Proceedings of the 2008 Annual International Conference of the Design History Society, (ed) F. Hackne, J. Glynne, V. Minto, 2-10. Falmouth: Universal Publications.

Latour, B. (2010): On the Modern Cult of the Factish Gods. Durham NC:

Duke University Press.

Latour, B. (2012): The Whole is Always Smaller Than Its Parts. A Digital Test of Gabriel Tarde's Monads, in British Journal of Sociology 63(4): 591-615.

Latour, B. (2013a): What's the story? Organizing as a Mode of Existence, in Organization and Organizing: Materiality, Agency, and Discourse, (ed) D. Robichaud and F. Cooren. New York: Routledge.

Latour, B. (2013): An Enquiry into Modes of Existence. An Anthropology of the Moderns. Cambridge, MA: Harvard University Press.

Latour, B. (2014): What is the Style of Matters of Concern, in (ed) Nicholas Gaskill & A.J. Nocek The Lure of Whitehead. University of Minnesota Press, pp. 92-127.

Latour, B. (2017): Facing Gaia. Eight Lectures on the New Climatic Regime. Cambridge, MA: Polity Press.

Latour, B., Woolgar, S. (1979): Laboratory Life. The Construction of Scientific Facts. Beverly Hills, CA: Sage Publications.

László, E. (1974): Preface, in Perspectives on General System Theory, by L. von Bertalanffy, (ed) E. Taschdjian. New York: George Braziller.

Lavie, D. (2008): The Competitive Advantage of Interconnected Firms, in Wankel (2008), Vol 1, 324-333.

Law, J. (1992): Notes on the theory of the actor-network: Ordering, strategy, and heterogeneity. Systems Practice, 5(4), 379-393.

Law, J. (1999): After ANT: Complexity, naming and topology. In J. Law & J. Hassard (Eds.), Actor network theory and after (pp. 1-14). Blackwell.

Law, J. (1987): Technology and heterogeneous engineering: The case of Portuguese expansion. In W. Bijker, T. Hughes, & T. Pinch (Eds.), The social construction of technological systems: New directions in the sociology and history of technology (pp. 111-134). MIT Press.

Law, J. (2019): Material Semiotics. heterogeneities.net/publications/Law2019MaterialSemiotics.pdf

Levin, M. (2022): Technological Approach to Mind Everywhere: An Experimentally-Grounded Framework for Understanding Diverse Bodies and Minds, in Front. Syst. Neurosci. 16:768201. doi: 10.3389/fnsys.2022.768201

Lewis, T. G. (2009): Network Science. Theory and Applications. New York:

Wiley & Sons.

Luhmann, N. (1971): Sinn als Grundbegriff der Soziologie, in Habermas/ Luhmann 1971.

Luhmann, N. (1975): Macht. Fink: Stuttgart.

Luhmann, N. (1977): Interpenetration – Zum Verhältnis personal und sozialer und sozialer Systeme, in Zeitschrift für Soziologie, Jg. 6. Heft 1. Jan.: 62-76.

Luhmann, N. (1982): The World Society as a Social System, in International Journal of General Systems 8(3): 131–138.

Luhmann, N. (1988): Wie ist Bewusstsein an Kommunikation beteiligt?, in Gumbrecht/Pfeiffer 1988.

Luhmann, N. 1988: Power, 2nd ed, Stuttgart: Franke.

Luhmann, N. (1989): Ecological Communication. Chicago: University of Chicago Press.

Luhmann, N. 1990): Essays on Self-Reference. New York: Colombia University Press.

Luhmann, N. (1992): What is Communication?, in Communication Theory, 2(3): 251-259.

Luhmann, N. (1993): Zeichen als Form, in Baecker 1993.

Luhmann, N. (1995): Social Systems. Stanford, CA: Stanford University Press.

Luhmann, N. (2008): Die Moral der Gesellschaft. Frankfurt a.M.: Suhrkamp.

Luhmann, N. (2012): Theory of Society V. 1. Stanford: Stanford University Press.

Luhmann, N. (2013): Theory of Society V. 2. Stanford: Stanford University Press.

Lupton, D. (2014a): Self-tracking Cultures: Towards a Sociology of Personal Informatics. Paper delivered at the OzHCI Conference 2014. (https://simplysociology.files.wordpress.com/2014/09/self-tracking-cultures-ozchi-conference-paper.pdf).

Lupton, D. (2014b): Self-Tracking Modes: Reflexive Self-Monitoring and Data Practices (August 19, 2014). Available at SSRN: http://ssrn.com/abstract=2483549 or http://dx.doi.org/10.2139/ssrn.2483549.

Lupton, D. (2014c): Apps as Artefacts: Towards a Critical Perspective on Mobile Health and Medical Apps. Societies 4, 606-622.

Lyotard, J.-F. (1984): The Postmodern Condition. A Report on Knowledge. University of Minnesota Press.

Malapi-Nelson, A. (2017): The Nature of the Machine and the Collapse of Cybernetics: A Transhumanist Lesson for Emerging Technologies. Palgrave Macmillan.

Mann, S. F., Pain, R,. Kirchhoff, M. D. (2022): Free Energy: A User's Guide. Biology & Philosophy. 37:33.

Marc, E., Picard, D. 1991: Bateson, Watzlawick und die Schule von Palo Alto. Frankfurt a.M.: Athenaeum Program.

Maturana, H. R. (1987): Cognition, in: Schmidt 1987.

Maturana, H. R., Varela, F. J. (1980): Autopoiesis and Cognition: The Realization of the Living. Boston Studies, in the Philosophy of Science.

Maturana, H. R., Varela, F. J (1987): The tree of Knowledge. Biological Basis of Human Understanding. Boston: Shambala.

Malapi-Nelson, A. (2017): The Nature of the Machine and the Collapse of Cybernetics, Palgrave Macmillan Cham.

Manicka, S., Levin, M. (2022): Minimal Developmental Computation: A Causal Network Approach to Understand Morphogenetic Pattern Formation, in Entropy 2022, 24, 107. https://doi.org/10.3390/e24010107

Miller, J. G., Miller J. L. (1990): The Nature of Life, in Systems Research and Behavioral Science 35(3):157-163.

Moltmann, J. (1985): Gott in der Schöpfung. Ökologische Schöpfungslehre, München: Kaiser.

Muller, A., Shipton, C., Clarkson, C. (2022): Stone Toolmaking Difficulty and the Evolution of Hominin Technological Skills, in Sci Rep 12, 5883. https://doi.org/10.1038/s41598-022-09914-2.

Mujis, D., Ainscow, M., Chapman, C., West, M. (2011): Collaboration and Networking, in Education. Dordrecht: Springer.

Nassehi, A. 1992: Wie wirklich sind Systeme? Zum ontologischen und epistemologischen Status von Luhmanns Theorie selbstreferentieller Systeme, in: Krawietz/Welker 1992.

OECD (2007): Giving Knowledge for Free: The Emergence of Open Educational Resources. Center for Educational Research and Innovation. http://www.oecd.org/edu/ceri/38654317.pdf

Open Educational Resources (2013): The William and Flora Hewlett Foundation. http://www.hewlett.org/programs/education/open-educational-resources.

Orban, G. A., Caruana, F. (2014): The Neural Basis of Human Tool-Use, in Front. Psychol., 09 April, Sec. Cognition, Volume 5 – 2014. https://doi.org/10.3389/fpsyg.2014.00310.

Parker, G., Choudary, S. P., Van Alstyne, M.W. (2016): Platform Revolution: How Networked Markets are Transforming the Economy – And How to Make them Work For You. New York: Norton.

Peirce, C. S. 1983: Phenomenon and logic of signs. Frankfurt a.M.: Suhrkamp.

Podolny, J. M./Page, K. L. (1998): "Network Forms of Organization," in: Annual Review of Sociology, 24:57-76.

Peirce, Ch. S. (1991): Peirce on Signs: Writings on Semiotic by Charles Sanders Peirce. J. Hoopes (ed). University of North Carolina Press.

Preda, A. (1999): The Turn to Things: Arguments for a Sociological Theory of Things, in The Sociological Quarterly Vol. 40, No. 2 (Spring, 1999), pp. 347-366.

Principia Cybernetica Website: http://pespmc1.vub.ac.be/CYBSHIST.html.

Riegas, V., Vetter, Ch. (eds) (1990): Zur Biologie der Kognition. Ein Gespräch mit Humberto R. Maturana und Beiträge zur Diskussion seines Werkes. Frankfurt: Suhrkamp.

Roth, G. 1987: Autopoiese und Kognition: Die Theorie H. R. Maturanas und die Notwendigkeit ihrer Weiterentwicklung, in Schmidt 1987.

Rowlands, M. J. (2010): The New Science of the Mind: From Extended Mind to Embodied Phenomenology. Cambridge MA: MIT Press.

Saz-Carranza, A., Ospina, S., Vernis, A. (2008): Leadership in Interorganizational Networks, in Wankel (2008), Vol. 2, 291-300.

Schmidt, S. J. (ed) (1987): Der Diskurs des Radikalen Konstruktivismus. Frankfurt: Suhrkamp.

Sebeok, Th. (1991): A Sign is Just a Sign, Bloomington: Indiana University Press.

Sebeok, Th. (1994): Signs. An Introduction to Semiotics. Toronto: University of Toronto Press.

Schiewek, W. (1992): On the neglected connection between 'symbolic generalization' and 'language' in social systems theory, in: Krawietz / Welker 1992.

Schmidt, S. J. (ed) (1987): Der Diskurs des radikalen Konstruktivismus.

Frankfurt: Suhrkamp.

Schmidt, S. J. (1994): Kognitive Autonomie und soziale Orientierung. Konstruktivitische Bermerkungen zum Zusammenhang von Kognition, Kommunikation, Medien und Kultur. Frankfurt a.M.: Suhrkamp.

Schuilenburg, M., Peeters, R. (eds) (2020): The Algorithmic Society. Technology, Power, and Knowledge. Routledge: London.

Shannon, C., Weaver, W. (1949): The Mathematical Theory of Communication, Urbanna: University of Illinois Press.

Shirky, C. (2008): Here Comes Everybody. How Change Happens When People Come Together. New York: Penguin Press.

Schneider, D. (2014): Personal Learning Environment. Edutech Wiki http://edutechwiki.unige.ch/en/Personal_learning_environment (retrieved Jan. 2016)

Siemens, G. (2005): Connectivism: A Learning Theory for the Digital Age, in International Journal of Instructional Technology and Distance Learning. Vol. 12. No. 1 Jan.

Simon, H. (2019): The Sciences of the Artificial. MIT Press.

Spencer Brown, G. (1972): Laws of Form. New York: Julian Press.

Stadler, M., Kruse, P. (1992): Zur Emergenz psychischer Qualitäten. Das psychophysische Problem im Lichte der Selbstorganisationstheorie, in Krohn/Küppers 1992.

Steffen, W., Richardson, K., Rockström, J. et al. (2020): The Emergence and Evolution of Earth System Science. in Nat Rev Earth Environ 1, 54–63.

Steier, F. (ed) (1991): Research and Reflexivity, London: Sage.

Surowiecki, J. (2004): The Wisdom of Crowds: Why the Many Are Smarter Than the Few and How Collective Wisdom Shapes Business, Economies, Societies and Nations. New York: Randomhouse.

Thomas, G. (1992): World as Relative Unity and as Last Horizon, in: Krawietz/Welker 1992.

Thorelli, H. B. (1986): Networks: Between Markets and Hierarchies, in Strategic Management Journal, V. 7, 1 (Jan.-Feb.), 37-51.

Varela, F. J., Maturana, H. R., Uribe, R. (1974): Autopoiesis: The Organization of Living Systems, its Characterization and a Model, in BioSystems 5, 187-196.

Wankel, C. (ed) (2008): 21st Century Management: A Reference Handbook.

2 Vols. UK: Sage Publications.

Watzlawick, P. (1977): How Real is Real?. New York: Vintage.

Weick, K. E. (1979): The Social Psychology of Organizing (2nd ed). New York: McGraw-Hill.

Weick, K. E. (1995): Sensemaking in Organizations. Foundations for Organizational Science. London: Sage Publications.

Welker, M. (ed.). (1985): Theologie und funktionale Systemtheorie. Luhmanns Religionssoziologie in theologischer Diskussion. Frankfurt: Suhrkamp.

Wiener, N. (1948): Cybernetics: or Control and Communication in the Animal and the Machine, New York: John Wiley & Sons.

Willke, H. (2004): Systemtheorie I. Eine Einführung in die Grundprobleme. 7th Edition, Stuttgart (UTB)

Willke, H. (1999): Systemtheorie II: Interventionstheorie. 3rd Edition, Stuttgart (UTB)

Willke, H. (1995): Systemtheorie III: Steuerungstheorie. 3rd Edition, Stuttgart (UTB)

Williams, I. (2020): Contemporary Applications of Actor Network Theory. Palgrave Macmillan: Singapore.

Wittgenstein, L. (1922): Tractatus logico-philosophicus. Kegan Paul: London.

Wittgenstein, L. (1953): Philosophical Investigations. Blackwell Publishing: UK.

Wolfram S. (2002); A New Kind of Science, Wolfram Media; and online https://www.wolframscience.com/nks/.

Wüthrich, H. A., Osmetz, D., Kaduk, S. (2009): Musterbrecher – Führung neu leben. Wiesbaden: Gabler.

www.ingramcontent.com/pod-product-compliance
Lightning Source LLC
LaVergne TN
LVHW102009210926
848483LV00010B/234